The International
BIOTECHNOLOGY
Handbook

Facts On File Publications
New York, New York ● Oxford, England

The International Biotechnology Handbook
First edition 1988

Published in the United States of America by
Facts On File, Inc.,
460 Park Avenue South, New York,
N.Y. 10016

Library of Congress Cataloging-in-Publication Data

The International biotechnology handbook
  Bibliography: p.
  Includes index
  1. Biotechnology—Handbooks, manuals, etc.
  2. Biotechnology industries—Handbooks, manuals, etc.
  I. Hack, Ron.
  TP248.2.I578   1988   660'.6   87-36406
ISBN 0-8160-1925-8

Printed in Great Britain
10 9 8 7 6 5 4 3 2 1

Contents

Foreword

This handbook presents a comprehensive picture of the international biotechnology industry in the mid 1980s. The industry has had its fair share of teething troubles and financial disappointments, but prospects for the new biotechniques to make a real contribution to industry and agriculture in the next decade are beginning to look up.

Although the book describes the scientific progress being made in biotechnology, the real accent is on practical economic issues – products on the market, major companies, and the standing of the industries in the main developed countries of the free world (USA, Japan and Western Europe). Other practical issues covered are the state of government regulation and patenting with regard to biotechnological products.

The book is in four parts:

Part One gives an introduction to biotechnology for the layman who may have only a vague idea, if any at all, of what this new science means. Chapter One briefly traces the historical developments which have led to man's imperfect but growing understanding of microbiological processes in the late twentieth century. Chapter Two follows this by outlining the main industrial products and processes with which biotechnology is involved today – again in simple terms for the non-expert reader.

Chapter Three examines in more detail (and necessarily in a more complex language) the products and processes which are the focus of current research in international biotechnology laboratories.

Part Two turns the spotlight on the national development of biotechnology in the major countries which are most likely to be the moving force in the next decade. Individual chapters are devoted to the USA and to Japan, while Chapter Six summarises developments in Western Europe. These chapters also provide profiles of the major companies involved in biotechnology, whether they be small, research-based (or 'start-up') companies or the large industrial combines best able to exploit commercialisation of biotechnology.

Part Three, in contrast, examines the broader issues of the biotechnology industry – which confronts researchers and commercial enterprises in all countries – with the main focus on regulatory issues (Chapter Seven) and patents (Chapter Eight). This part of the book concludes with a chapter on market trends and prospects, a chapter which is a natural sequel to Chapter Three on current research. An individual chapter (Nine) is devoted entirely to the impact on medicine of biotechnology, because this is the sector of the economy which is already beginning to feel the impact of new products and processes based on this 'new' science.

Finally, Part Four is a reference section containing seven different sources of information on biotechnology, including a bibliography, press section, databases and, in line with the economic stress of the book, a section on current market surveys. The source lists are introduced by a guide to finding out more about biotechnology using these different sources.

Each chapter of the book has been researched and written by an expert contributor who specialises within the particular field concerned, whether it be market research, scientific research or information science (see List of Contributors). This means that each chapter has its own 'flavour', or a style suited to the topic being covered, whether industrial, scientific or commercial. The 'national' chapters (Four, Five and Six) have been written by contributors based in those countries, thus ensuring a difference of viewpoint in each chapter although the substance of the industrial coverage is the same throughout.

The format of the book does mean that some repetition of fact has been unavoidable. However, the reader will find that this repetition does serve to emphasise the most important aspects of biotechnology in the mid 1980s.

List of contributors

Contributors to The International Biotechnology Handbook include Ron Hack, a technical editor/journalist (*historical development, industrial applications*); Dr. Tazewell Wilson, a biotechnology specialist with consultants Korda & Co. (*current research trends, market trends and prospects*); Christopher Pleatsakis, a US market analyst based in California (*biotechnology industry in the USA*); Conje Hallstrom, a Tokyo-based economic and marketing consultant (*biotechnology in Japan*); David Tucker, a UK based researcher and writer specialising in the world health industry (*biotechnology industry in Western Europe*); Dr. M. G. Norton, Head of Biotechnology at the Warren Spring Laboratory (*regulatory issues*); R. S. Crespi of the British Technology Group (*patenting in biotechnology*); John Hodgson, a science writer and Editor, Trends in Biotechnology (*impact on medicine* and Dr. Anita Crafts-Lighty, Managing Director of BioCommerce Data Ltd. (*information resources for biotechnology*).

Part One
Biotechnology Today: An Overview

1 Chapter One

Historical Development of Biotechnology

This section is in no sense intended to be a comprehensive history of the development of biotechnology – that book is yet to be written and a fascinating work it will be. However, by roughly tracing the ways in which humans have made use of (or rather co-operated with) microbes over the centuries, this complex subject can be put into some kind of perspective. Biotechnology divides into three parts: the early 'magical period' where discoveries made by accident were incorporated into daily life simply by following rules of thumb; a middle period beginning around the seventeenth century, when scientific explanations for these simple processes began to be established and followed; and the modern period when microbes have become the basis of great industries and modern techniques have given us new and far-ranging powers over micro-organisms.

1.1 Origins

Human beings have been making microbes work for them for a very long time. Oldest of all the useful products perhaps is wine, possibly as old as eight thousand years. Simon van der Weele, one of the few historians of the subject (so far), put it neatly:

> "Wine must be as old as pottery. – What do you do with a pot? You fill it with berries. Like most humans you forget one pot and the berries ferment. The first person to consume the contents had once again discovered wine. Vinegar is even easier, you just forget one pot of wine."

1

Sadly, history does not record even one of the many 'discoverers' of wine (unless Noah can be given this honour). Similarly, nobody knows who first added yeast to bread to make it 'rise'. The first time is pretty certain to have happened by accident. We know where, and roughly when, and we can make a guess about how.

The Nile Valley was the place and the time about five thousand years ago (during the reign of the Pharaohs). Certainly, the later Egyptians had bread and the Old Testament, somewhat later, is clear about the difference between 'leavened' and 'unleavened' bread. The difference, familiar to everyone who has tried baking, is in the lightness and texture that yeast gives to bread. When yeast is added to the dough mixture it begins to multiply (because, of course, it is alive). You must give it a little time to rest or 'prove'. As the yeast grows, or 'ferments', it gives off the harmless gas carbon dioxide. This is trapped as bubbles in the dough and results in the lightness and airiness that we know as the texture of bread.

Yeasts (there are many kinds) are naturally occurring one-celled microbes, and they are harmless and fairly common, floating around in the air. On the first occasion (in reality, like the wine, it must have happened many times in different places) a 'wild' yeast just happened to get into the bread mixture, perhaps from soured milk in which some yeast had already multiplied.

That batch of loaves must have tasted particularly good, and in a superstitious way maybe the baker saved a bit of that 'special' dough and added it to the next batch. Without really knowing why, he was 'seeding' the mixture with enough live yeast to multiply and aerate the next batch. This process went on down the centuries. Until pure yeast became commercially available, each batch was seeded with a little dough from the last.

The other great quality of yeast must have been discovered about the same time. This microbe lives on the sugars that are present in various quantities in most vegetable matter. In the right conditions it will grow and multiply incredibly quickly – this is known as 'fermenting'. As it grows, as well as giving off the carbon dioxide that makes such a difference to bread, it produces another much more interesting substance – alcohol.

So brewing started, first as yet another household chore, and then semi-industrially, in the back sheds of pubs and taverns, and still later in specialised factories – breweries. As well as beer, these establishments sold another product: 'brewer's yeast', skimmed off their beer vats. And this in turn was sold to bakers and householders.

The Egyptians were first again in this field. They produced a kind of beer from fermented cereals, but other primitive cultures produced their own versions. The materials fermented varied from asses' milk to various grains, barley and oats (the beer of many peoples and the 'ale' of the Vikings). Wine (from grapes, of course) was well known to the Greeks and Romans and spread across the world under their influence.

A number of other primitive biological processes were known to the ancient world. Notable among them are the manufacture of cheese and leather. Cheese is a fascinating substance. Originally it must have been just a way of conserving surplus milk, but later developed into the staple food (and delicacy) that we know today.

The key process in making cheese is coagulating (or solidifying) the liquid milk into the cheese solids (or 'curds'). Some unknown genius back in the mists of time must either have used a calf's stomach as a container for milk or gone through the remarkable thought process 'Calves live on milk alone – there must be something in their stomachs that turns the milk into a solid food.'

There is such a something – we now know that it is an enzyme called chymosin, about which more will be said. For many centuries this substance (then known as 'rennet') could be obtained only by boiling down calves' stomachs. Using rennet from this unlikely and unsavoury source, the entire cheese-making industry that we know today grew up.

More likely as a container than the calves stomach perhaps, was a leather bag or bottle. Tanned leather was an incredibly useful material during this period, not only for harness, straps and boots (for which it is still unsurpassed), but for bottles and goblets, clothes and armour. Other enzymes were used by the tanner to break down the stiffness of raw leather and turn it into a pliable and durable material. Like all the other users of microbes he had to work carefully and methodically, making sure, by meticulous ritual, that only the right bugs were present in his baths and vats. For there were, and still are, microbes that rot leather, make beer sour and turn wine into vinegar.

It is certain that the craftsmen and housewives of ancient history had no clear knowledge of why these processes worked. At best, there was a sort of semi-magical explanation for them. Today, with a fuller knowledge of just how complex these processes are, scientists have a mature admiration for so-called 'primitive' people who could develop, by whatever method, a fairly reliable means of producing good bread and cheese, beer, wine and supple leather.

Now, with advances in biochemistry and genetics, accurate instruments and powerful microscopes, we have more or less exact data on what happens in a vat of beer or a round of cheese. We can now control the processes more precisely and avoid some of the pitfalls in manufacture. With a few exceptions, however, we still make our bread and beer, and our cheese and leather by the methods developed centuries ago.

1.2 The Scientific Age

What we might call 'the age of innocence', during which mankind used microbes without really knowing what they were, lasted until about 1876, when Louis Pasteur identified some unwanted microbes that were spoiling the fermentation of beer. Beer and wine had gone bad before this, of course, but mishaps had been explained by the same kind of 'magic' as was thought responsible for the proper working of the process. When your beer went off, it was because of witchcraft, the evil eye, or even a change in the weather (which could have been correct if it radically changed the temperature at which the brew was made).

Microbes themselves had been seen and identified some two hundred years before by the Dutchman Antoni van Leeuwenhoek (1680), who invented a primitive form of microscope. What exactly they did, however, remained a mystery until Pasteur put forward his 'germ theory' and identified a number of bacteria and their functions. He was asked by Napoleon III of France, in 1863, to look into the reasons why wine (even then an important French export) deteriorated on its way to the consumer.

Pasteur introduced three very simple improvements in the approach to fermentation. The first, a rather obvious one, is that unless the wine is kept, not only in clean utensils, but also excluded from the air (and the many stray yeasts that abound in it) it will quickly turn into something else – usually vinegar.

This approach to hygiene led to the second innovation, the process that bears his name—'pasteurisation'. In this, wine, beer or any other product, including milk, is heated and held at a temperature just below its boiling point. By killing off most of the bacteria contained in the liquid, the onset of unwanted fermentations (which make beer or milk go sour) can be greatly postponed. We should distinguish this process, of course, from 'sterilisation', in which the liquid is raised above its boiling point for rather longer. This can kill many more bacteria, but often has a harmful effect on the taste and structure of the product.

Pasteur's third innovation was to recognise the importance of oxygen to fermentation. When yeast is grown without sufficient air, it can only partly convert the available sugar into alcohol. Careful aeration of the fermentation vessels thus leads to stronger wine, and controlled stirring and aeration remain important features of modern fermentation techniques. There is, of course, a limit to this process. Alcohol, as far as the yeast is concerned, is a 'waste product', and when the concentration of alcohol in the wine or beer reaches around 18%, the yeast is poisoned. There is thus a limit on the alcoholic strength of naturally fermented liquor. Stronger liquors are, of course, obtained by distillation of fermentation products.

Unconnected with the work of Pasteur, but contributing greatly to our knowledge of genetics, was the work of the monk Gregor Mendel in the mid 1800s. Working in a garden, with several varieties of peas, he established precise rules governing the transmission of inherited characteristics. Mendel's contribution was an elaborate structure of 'rules' from which plant breeders eventually developed their methods of 'crossing' and hybridisation, and stockbreeders wrote their 'stud books'. What happened was now fairly clear. How it happened, the mechanisms involved in genetic transmission, was not discovered until the turn of this century.

In parallel with these research efforts was the beginning of the industrialisation of fermentation processes, the practical side of biotechnology. Breweries were now big business, as were distilleries, and the production of baker's yeast was taken out of the backstreet shop with the establishment of specialist factories.

The pressures of World War I also encouraged the production of industrial chemicals by fermentation. The best example of this is the production of acetone, an essential ingredient in cordite, for explosives. In 1914, Weizmann introduced the microbiological manufacture of acetone and butanol in the UK. As in the case of penicillin in World War II, the USA quickly contributed its facilities for large-scale production. Contemporary photographs show rows of fifty thousand gallon tank fermenters, the largest seen until then.

The foundations of modern biotechnology were laid in the first half of this century. The spectacular laboratory discoveries of recent years, the 'gene boutiques', have no real future if they remain curiosities limited to tiny demonstration quantities. The techniques of large-scale commercial production and the accumulation over the years of expertise in fermentation and extraction have a very important part to play in the development of biotechnology into a real industry.

1.3   The New Era

In most fields of human endeavour, one can trace a progression of small developments and improvements, interspersed with occasional spectacular spurts. It is fashionable to call these 'breakthroughs' or 'quantum leaps', but in general these are only the outward indications of broad progress on many fronts.

The recent spectacular development of techniques of 'gene-splicing' (genetic engineering) are of such significance that it is usual nowadays to describe these processes as 'new biotechnology' to distinguish them from all that went before. Quite a lot, however, had gone on before, laying a foundation for this research and the industries that are growing around it.

Genetic research, applied to microbiology, had of course been in existence for many years before these new techniques were perfected. The practical side, however, had been much like stock-breeding, limited to careful selection of the best strains. Methods were primitive and painstaking, from cultures were made anything and everything, using soil samples from exotic places, bits of plant and animal tissue, and sometimes the accidental intrusion of yeasts and moulds from the air. Cultures were made by 'seeding' (sowing your sample) on a surface of jellified broth (containing agar) in a round, flat glass dish (a 'petri dish'). You then left it in a suitable place and watched what happened. 'Colonies' develop where microbes touch the agar surface and these can then be isolated and examined for useful properties. Simple it may be, but spectacular discoveries (penicillin, for example) have been made in this way.

Microbes have the advantage, of course, of reproducing many times faster than plants or animals, so many generations can be examined in a relatively short time. This rapidity also provides an opportunity to exploit variations that either occurred naturally or could be produced in relatively simple ways (irradiation, for instance, and by certain chemicals).

A 'mutation' or 'sport' (well known in the breeding of animals) is an animal or plant that has a characteristic not apparently given to it by its parents. The results are entirely a matter of chance, some mutations are successful, some not. Such chance mutations are a key factor in Charles Darwin's theory of 'natural selection'. According to this, organisms with the more successful mutations (longer necks, specialised diets) had a small competitive edge over their fellows and prospered, gradually changing, or evolving, their species.

It had been known for many years, since the work of Mendel in the nineteeth century, that every life form contains within its cells the 'blueprint' for reproducing itself. Mendel laid out a structure of rules governing the transmission of this information, but the mechanisms involved had to wait for the more sophisti-cated techniques (and more powerful microscopes) available in this century.

This research began to suggest that the information was somehow encoded in the 'chromosomes' – tiny bodies visible as rod shapes in the nucleus of each cell. Every organism has a complement of these in varying numbers. Humans have 26 pairs, a crayfish has over 200, bacteria have only one, and no nucleus. Later studies showed that the chromosomes (the name simply means 'coloured shape') are themselves made up of even tinier units called genes. It was the patterns contained in these genes that in some way enabled the organism to replicate itself or 'breed true'.

It was found that the genes were composed of a complex chemical compound known as DNA (deoxyribonucleic acid). It was not, however, until as recently as 1953 that two researchers working at Oxford, Francis Crick and George Watson, finally uncovered the precise shape and function of the DNA molecule. This was the now famous 'double helix' – roughly the shape of a spiral staircase.

When the cell reproduces, the rungs of this staircase split down the middle of each tread and the two spirals that remain (with the half treads) act as a template for the formation of new comple-mentary molecules – exact copies of the original, parent DNA. The threads of the staircase are composed of strings of units known as 'nucleotides', different kinds of nucleic acid. There are only four different types, but the 'staircase' is extremely long and the sequence of nucleotide 'bases' can easily convey a great deal of information. The best analogy here is the Morse Code. With just two different symbols (the dot and the dash), any message,

7

including, if you like, the works of Shakespeare, can be encoded. The arrangement of these nucleic acids, in specific sequences, carries all the detailed information necessary for the construction of a new organism – the genetic code.

GENETIC ENGINEERING

This slightly misleading term covers the recent development of techniques concerned with producing what are effectively controlled mutations – deliberate evolution.

The techniques, often colourfully called 'gene-splicing', are methods of constructively rearranging the genetic code to produce an organism with new, desirable characteristics. In the case of a simple microbe, this might involve introducing the ability to produce a certain chemical. Somewhere on the DNA molecule is the information ('genes') concerned with this desired quality or product. The 'engineering' is concerned with 'cutting out' that part of the string, and joining or grafting this into another organism.

This is not, of course, done with a knife, but chemically by using special enzymes. The discovery of this way to use 'restriction enzymes', by Stanley Cohen and Herbert Boyer in the early 1970s, supplied a tool that would cut the long DNA molecule into a fixed number of defined fragments. The enzyme does this by recognising and attacking certain specific nucleotide sequences in the DNA chain. There are now well over 300 of these enzymes in general use, each 'tuned' to a different sequence. 'Splicing' is done with another set of enzymes called 'litigation' enzymes, which can stick the fragments back together.

All this talk of 'cutting' and 'splicing' perhaps suggests delicate surgery, but it must be remembered that all these functions are in practice carried out chemically, in solution. The molecular biologist/genetic engineer often ends up with a 'soup' composed of fragments of the genetic code. Somewhere in there is the coding for the desired characteristic, but, unneeded sequences will also be present. There is a number of techniques that have been developed to help in the selection of those fragments he wishes to process.

If a DNA sample is placed on a plate of gel under appropriate chemical conditions and a weak electric current is passed across the plate, the fragments will start to 'migrate' in the direction of the positive pole. The smaller ones move faster, and after several hours they are all arranged in a neat line from the shortest to the longest.

Electrophoretic separation can be used to determine the sequence of 'bases' in a DNA fragment or to separate specific molecules from one another. The DNA synthesiser (or 'gene machine') is a fairly recent development which promises greatly to expand the scope of the gene-splicer. Since there are only four components (however complicated their permutations and combinations) to the DNA string, it has been possible to construct an apparatus that can create a specific sequence of DNA, a synthetic gene.

When a desired fragment is identified or synthesised it can be re-inserted into a 'host' organism. Genetic engineers tend to choose micro-organisms with thoroughly known characteristics. Esherichia coli, a common intestinal bacterium, is a favourite as is ordinary baker's yeast (*Sacchromyces cerevisiae*). The resulting 'engineered' organism is then 'cloned' from a single cell grown and multiplied in the usual way and if the process is successful it will breed true – a new organism with an added ability engineered in. To get DNA into a 'host', a 'vector' must be used. Usually this is a small piece of DNA which is attached to the new DNA and when inserted 'recombines' again with the chromosmal DNA using natural processes usually involved in the cell's reproduction.

The process has two distinct aims. It can simply be used to increase the production capability of the microbe concerned. This, known as 'amplification', consists of increasing the number of gene sequences devoted to the production of the desired substance. Alternatively, an entirely new production ability can be grafted in. Genetic material from literally any source, plant or animal, can be inserted into a microbe whose growth and fermentation characteristics are thoroughly well known.

PRACTICAL APPLICATIONS

To take a recent example, the enzyme chymosin (the coagulating agent rennet that is essential to the cheese industry) exists in nature only in the stomachs of calves. Thus, like insulin (which is found in the pancreas of pigs and cattle), it was formerly available only by extraction from slaughterhouse products.

By using genetic material taken from a calf's stomach, researchers were able to isolate the ability to produce chymosin and then graft this gene on to a well-known and well-documented microbe such as yeast and E coli. The result is a product that is identical in every respect to the original, and the new product can be produced to match the demands of the cheese market – a much more satisfactory state of affairs than reliance on the availability of calf stomachs, which depends on the market for veal.

Any new substance, particularly if it is for human consumption or treatment, has a long development cycle, followed by an equally long test period. Insulin, for controlling diabetes, was the first drug to be engineered in this way. It was licensed for human use in 1982. A number of other products are currently in the process of development. Interferon and growth hormone are also now on sale in several countries. Although there is a requirement for rigorous clinical trials of new therapeutic products, the con-tinuing need for better treatment methods provides a strong pressure to adopt any which prove effective. This trend is less marked for chemical, food and agricultural products where effective but perhaps less efficient or economical processes already exist. For example, rennet produced by recombinant DNA methods is not yet on sale anywhere though successful trials in cheese making have been reported since the early 1980s. Many complicated commercial factors are involved in this, including the size of the market and the validity of patents, but chymosin is also a good example of some of the complexities of genetic engineering.

Putting a gene into a new organism so that it can be replicated doesn't necessarily mean that it will work. Genes direct the synthesis of 'proteins', molecules which are made up of 'amino acids'. Every three DNA bases specify one amino acid of a protein (the key to the genetic code), but the process by which genes are 'transcribed' into proteins is a very complicated one involving another kind of nucleic acid (ribonucleic acid or RNA) and other proteins which are chemical catalysts (enzymes). To complicate matters further, DNA contains portions which are not 'genes' in the sense of coding for proteins but are involved in the 'expression' of genes for example by providing places for the enzymes involved in replication to bind. Such 'promoters' and 'enhancers' can be critical to a gene's function, so they sometimes have to be put in along with the 'cloned' gene, or the gene has to be recombined at just the right spot in the chromosome or it will not work properly. Proteins are not always assembled quite right by the new host and may need to be modified by other cellular enzymes before they work properly. Chymosin when made in E

10

coli is often found in insoluble granules because of such problems, which makes its extraction and purification more difficult and costly.

At this stage it is appropriate to review our terminology. The chymosin produced from yeast is, by definition, genetically identical to the enzyme produced in a calf's stomach. Is it then a 'natural product' or should it labour under the description 'synthetic'? With the emotional reactions surrounding these terms nowadays, we must tread with some care. The general feeling among scientists is that a substance produced in this way has every right to be regarded as natural.

A similar dilemma will be encountered in a later chapter, covering the medical aspects of biotechnology. Therapeutic products formerly produced from extracts, particularly growth hormone, will make a spectacular contribution to the treatment of such conditions as dwarfism, and may give the medical profession much greater control over human height in general. They may also have uses in wound healing, though such applications are not well proven yet.

Until now, the only source has been recovery from human cadavers, a highly expensive and obviously limited source. A microbial method of production of such hormones will have a two-fold advantage. Not only will there be more of such hormone products available for treatment, but there will also be ample for the research which should precede such treatment. But human society is also a complex organism, easily upset by such advances. If the potential exists for ensuring that every child can grow to six foot (or even seven foot?), will this be the option that parents will choose? The dilemma is similar to that facing doctors in many countries in recent years, when it became possible to predict the sex of an unborn child. In societies which place a disproportionate value on male children, will enough females get to be born?

ISSUES FOR THE FUTURE

So far, we have covered only the engineering of products available from other sources, albeit with difficulty. What if there is no limit on the kind of substance or organism that can be produced? It should be remembered that these techniques are 'open-ended' in the widest sense of that term. Researchers at this time cannot, with any precision, see bounds to what could eventually be possible.

11

It seems likely that *any* substance produced or secreted by an animal or plant can be reproduced eventually in this way, perhaps more conveniently, more safely, more economically. The almost limitless potential of biotechnology in food, energy and medical science will be discussed more fully in a later chapter. Achieving such a potential will depend on the establishment of an effective infrastructure for commercial exploitation of the technology, but this may in itself lead to new problems of regulation and ethics.

In its potential long-term effects on society, there is more than a hint of parallel between biotechnology and microelectronics. This leads us to beware the kind of 'hype' that was fashionable in the early days of the so-called 'computer revolution'. Do you remember when the computer was going to change radically every aspect of human life? Let us inject here a word of caution from one of the leading researchers in the microbiology field.

> "However complex the computer may eventually become, every part of it will still be produced by human ingenuity and so will be analysable, controllable and repairable by human agency. With biotechnology, one is dealing with the complexity of Nature herself, many orders of magnitude more complicated than anything Man has yet created. The potential to change the world is certainly there, what is questionable is simply our ability to realise it in full – and to control it."

Bearing this in mind, one could still say that the potential to change human society implicit in biotechnology is at least as great as that in the computer, possibly many times more. Many informed scientists do say just this. The time scale, however, stretches over the next century. In this book, we shall endeavour to take a view of the emerging field of biotechnology as it appears in the mid 1980s, and consider some of the opportunities and challenges it now presents.

2 | Chapter Two
Industrial Applications

2.1 Introduction

As we saw in the last chapter, genetic engineering is only a small, though spectacular, facet of biotechnology. A powerful and highly developed industry has already grown up using the more traditional microbial techniques. Fermentation and cell culture today are carried on in huge vessels, 150 cubic metres and more, using highly developed computer control of temperature, aeration, and stirring to give the optimum conditions for production. Careful selection of production strains of microbes and improved methods of extraction and purification have increased yields many times over the last fifty or so years.

Today, these traditional methods are used to produce yeast and alcohol, enzymes, antibiotics, vaccines (and drugs of many kinds including steroids and hormones) and such basic materials for the food and other industries as citric acid, gluconic acid, lactic acid and many kinds of amino acids. Even without the recent advances in genetic engineering, the biotechnology industry would still represent an impressive force. With these new techniques, as we have seen, the limits are not readily discernible.

Let us, however, examine the principal components of this industry as they exist today, and leave speculation about developments tomorrow to a later chapter. There are three main sectors: yeast, enzymes and antibiotics, with a number of smaller (but potentially very important) divisions including pharmaceuticals, diagnostics, effluent treatment, energy, and agriculture.

2.2 Yeast

This, the earliest of Man's microbial servants, has become the basis of several important industrial complexes. The supply of fresh yeast to bakers all over the developed world is big and continuing business, on both a national and international scale.

The conditions under which the fermentation takes place can be adjusted to favour either the production of yeast itself or alcohol, though both are produced in quantity. In the present economic structure of Europe, yet another lake of alcohol is not desirable, but in many parts of the world, Brazil for instance, the fermentation of cane sugar to ethanol for fuel use is providing a viable alternative to petroleum. In the long term, with changing conditions, this could become an important strategic energy source. Though sugar cane will not grow in temperate climates, sugar beet has proved a reliable alternative, and experiments suggest that the Jerusalem artichoke may be an even better one.

We shall return to such speculation in a later chapter. In the meantime the highly developed yeast industry is an important reservoir of the traditional techniques of fermentation and extraction on an industrial scale. As we have seen, the new recombinant DNA techniques depend on the use of well-known and tried production microbes if they are to make the transition from laboratory curiosity to commercial production. For this reason many of the new engineered strains are devised from this familiar and well-documented production source.

As a developed yeast industry in Europe and the US has provided a base for new development in biotechnology, a similar industry exists in Japan based on the soya bean. The fermentation of this raw material into curds and sauce provides an important part of the Japanese diet, and with strong government help, this industry will no doubt play an important role on the international stage in the near future.

2.3 Enzymes

Enzyme production is another well-developed sector. No one goes out to the supermarket to buy a pound of enzymes, but many of the products on the shelves there have these important ingredients somewhere in their production cycle. Every cell has hundreds of enzymes. They are most important to the cell, and to us, because they are the catalysts which the cell uses to make things happen. Extracted, they can be used by us to start, stop or control many kinds of processes.

Enzymes make a chemical reaction occur faster or more efficiently without themselves being changed. In industry, particularly the chemical industry, the use of synthetic chemicals as catalysts is widespread, although their exact mechanism is still not always completely understood. Often their action is rather like a template, or pattern. Something in the surface construction of the enzyme allows the other molecules to combine or react.

As we have seen repeatedly, not knowing exactly how something works is no barrier to using it, and man has used enzymes for centuries. The word (Greek again) means 'in yeast', and the enzymes produced by yeast are by far the oldest of our microbial tools. By extracting and using enzymes, we are making use of the cell's own mechanisms for control. Most organic reactions, left to themselves, are very slow. Each cell, therefore, contains a battery of these enzyme 'control levers' with which it can affect its environment. They start (or stop) very simple but basic things happening around them.

Most of us are familiar with one such action, the protein-degrading enzyme (protease) contained in 'biological' detergents. The really stubborn stains in laundry are organic proteins: blood, food, wine and so on. When left to soak with the laundry, the enzyme breaks up these proteins, allowing them to be removed by normal washing action.

For many years a similar enzyme, amylase has been used in the textile industry for breaking down the starch which comprises the coating of yarn, preparatory to printing. Another important and familiar enzyme is rennin used for centuries to coagulate milk solids for cheese making, and other enzymes are used throughout the dairy industry.

In brewing, the next most important ingredient to the yeast, of which we have spoken before, is an enzyme (another kind of amylase) which breaks down starches in the barley into the sugars that the yeast can consume. This enzyme is produced naturally by 'malting' (gently heating the grain and encouraging it to sprout). This is an expensive process, however, and malting, particularly for light beers, is being by-passed by the addition of commercially produced enzymes. Others are used to help clarify the beer during production.

Starch breakdown is also the function of enzymes in the

Industrial Applications

production of glucose and high-fructose syrups from grain. This is used in many soft drinks today, and is a potential source of commercial alcohol in a different energy climate. Over the years, the use of enzymes has become essential to many industries, food, drink, drugs and textiles. An important industrial sector has grown up to supply these needs.

2.4 Antibiotics

The discovery of antibiotics was perhaps the most important single advance in the history of medicine, as a sober look at the situation before their discovery will confirm. The horrors of the 'septic ward' and infant mortality from such diseases as diphtheria and tuberculosis are happily in the past. Antibiotics are also a prime example of the importance of biotechnology to medicine. Their very working mechanisms are an example of the successful manipulation of micro-organisms and their manufacture and development has been an integral part of the story of bio-technology.

An antibiotic is basically a substance that, present in extremely small amounts, interferes specifically with the growth of micro-organisms. The important word 'specifically' in this definition refers to the fact that an antibiotic is precise in its effect on particular organisms, and also that it has little effect on the body of the patient or most other microbes. This distinguishes its action from that of an antiseptic, most of which have a drastic (unspecific) effect on bacteria, and can also seriously harm the patient by similarly affecting his own cells.

In 1929, Fleming published his now famous discovery, that a mould, Penicillium (a fairly common mould often found on stale bread) inhibited the growth of colonies of a dangerous bacterium, Staphylococcus, in its vicinity. Similar effects had been observed before, but Fleming was the first investigator correctly to conclude that it was some substance produced by the mould that was responsible for this 'antibiotic' activity.

Fleming demonstrated that this substance, that he named 'penicillin', was highly effective against Staphylococcus (the main organism responsible for septicaemia) and several other 'gram-positive' bacteria, though not against the numerous 'gram-negative' types. This classification was made by the nineteenth-century Danish scientist Hans Gram, who distinguished different varieties of bacteria by their reaction to a staining technique. It was later established that this difference is due to variations in the construction of the cell walls. It is these cell walls that are attacked by antibiotics such as penicillin. He also demonstrated that an active broth containing 'penicillin' was no more toxic to animals (and particularly their important and vulnerable white blood cells) than a normal harmless broth. Due to the instability of the material, however, Fleming abandoned his researches at this stage, and little happened for nine years.

In 1938, Florey and Chain, working in Oxford, made a systematic investigation of anti-microbial substances produced by micro-organisms. During this work, they uncovered the real significance of Fleming's work and developed it from a laboratory curiosity into a significant medical advance. They managed to purify small quantities of penicillin and were able to demonstrate its therapeutic usefulness on both animals and humans with spectacular results. It remained, however, difficult and expensive to produce the drug in any quantity.

Meanwhile, World War II had started and the problems of large-scale production were to be solved by two outside factors: the huge resources of American industry and the pressing need to produce this drug with such potential for treating battle casualties. The story of Anglo-American cooperation on this project is now well known. The chief result was the establishment of large-scale production by an adaptation of a traditional biotechnological technique, deep fermentation, still the principal method by which antibiotics are produced. This involves growing the micro-organisms in large tanks.

Penicillin itself has a limited effect on gram-negative bacteria, and worse, some bacteria have the ability to produce an enzyme 'penicillinase' that destroys penicillin. This ability, of course, did not appear magically. It was due to a simple 'evolutionary' effect in the presence of penicillin: those organisms which did produce the enzyme flourished in the absence of those which had been killed off. Thus began the race against resistant bacteria, which even now we are only just winning. The search for and development of new antibiotics continue and, equally important, the develop-ment of new hygienic techniques in hospitals to combat the effect of 'hospital infection' (resistant strains of bacteria) is still an important field.

The work on the development of new and more powerful antibiotics began with isolation and investigation of the complex penicillin molecule. This was found to consist of a molecule with side-chains that could be modified chemically, into a range of new products with new abilities including the ability to attack gram-negative and resistant bacteria. These modifications and further discoveries (such as that in 1953 of a new antibiotic source in the fungus Cephalosporium) gave rise to the huge range in antibiotics now available to the doctor. Antibiotics today are produced by a combination of fermentation processes and sophisticated chemical techniques, and their production and development forms an important part of the biotechnological industry.

2.5   Other Pharmaceuticals

Antibiotics have become so well established as mainstream pharmaceutical products that today they are usually excluded from many reviews of the opportunities for biotechnology in the pharmaceutical industry. Some of the most exciting prospects for new drugs come from a class of products called 'biologics'. Basically this means derived directly from a biological source, for example extracted from glands or blood.

BLOOD PRODUCTS

Blood products are an important class of biologic including such things as Factor VIII, a protein which clots blood and is lacking in haemophiliacs. Antibodies from blood are sometimes used to provide passive immunity and, of course, blood and plasma (what's left after all red and white blood cells are removed) are used directly in transfusions. Most blood is obtained from donors and recently the increased incidence of viral blood-borne diseases such as hepatitis and 'acquired immune deficiency syndrome' (AIDS) has led to risks of contamination, which is a major reason why products like Factor VIII are now targets for genetic engineering. Limits to supply and high products costs are also factors.

VACCINES

Passive immunisation with antibodies is not a satisfactory way to prevent most infectious diseases. Not enough immune donors are available, and their blood does not contain a lot of the specific antibodies needed unless they have just recovered. Also, antibody proteins cannot multiply themselves and are gradually broken down in the recipient. For these reasons, immunisation is usually done by injecting a weakened or killed preparation of the pathogenic microbe which does not make the patient sick but is enough like the disease-causing organism for the production of the correct antibodies to be stimulated. These are produced by white blood cells (B-cells) which 'learn' to make a particular form of antibody specific to a molecule present on the injected organism. When 'challenged' by a future infection with the real disease-causing strain, these cells will turn on antibody production and produce large amounts to protect against the invader.

It is not always possible to produce a safe, effective vaccine using killed cells or weakened strains, so biotechnology is being used to produce pure proteins which fool the immune system into making antibodies, and can be produced cheaply and safely using recombinant DNA techniques. Already, a new vaccine for AIDS is a key commercial target. For veterinary uses, better vaccines for diseases like rabies, feline leukaemia and foot and mouth may be possible.

DIAGNOSTICS

The production of antibodies in nature is exquisitely specific and a key component of the immune system. Each antibody recognises a tiny portion (the 'antigen') of the molecules making up the foreign ('antigenic') material. Normally, many different clones of B-cells are involved. Biotechnology, however, has provided a way to manufacture purer antibodies outside the body using the techniques of cell fusion. B-cells are fused with similar cancerous (myeloma) cells using a chemical, polyethyleneglycol (anti-freeze!) or the virus that causes glandular fever (Epstein-Barr virus). The resultant hybridoma cells initially contain double the correct number of chromosomes but rapidly divide, producing cells with various random assortments of chromosomes. Some of these will have the chromosome responsible for antibody production and the cancerous characteristic of unrestrained growth helpful if the cells are to be grown in culture. These clones produce a single, 'monoclonal' type of antibody which can be made in large amounts by growing the hybridoma cells in fermenters similar to those used for antibiotic-producing microbes. The antibodies so produced are not much use for vaccines, but they have proved very useful for diagnostic tests.

Until fairly recently most medical diagnosis was qualitative, involving physical symptoms and in some cases the culture of disease-causing bacteria or chemical analysis of blood samples (e.g. iron levels for anaemia). The specificity of antibodies has allowed a whole new type of blood test to be developed, but at first these used 'polyclonal' antibodies extracted from (usually animal) blood. These were variable in their performance and expensive to produce but nevertheless formed the basis of a number of clinical assays, for example for thyroid disorders. Monoclonal antibodies have revolutionised this technology because large amounts of very specific antibodies are now available cheaply. This has led to many more products including pregnancy and ovulation tests for humans and animals (particularly cows), assays for common infections such as strep throat and serious diseases such as AIDS and hepatitis. The same methods are being applied to cancer diagnosis too and have led to new tests which are so simple to perform they can be done in a doctor's office or at home. This will have a big effect on the cost of health care and the very way medicine is conducted.

2.6 Food Products and Agriculture

The pharmaceutical industry is characterised by products which are produced in small amounts (sometimes only grams per year, at most a few hundred tonnes) but have a very high value (sometimes thousands of $ per gram). Characteristically it spends a lot on research, and the high profit margin and patentable nature of the products make the development costs worthwhile despite the long lead times required for clinical testing and regulatory approval. The food industry, by contrast, deals in 'high volume/low value' products, typically with fairly low profit margins. As modern biotechnology depends heavily on costly research, there is perhaps less room for impact here in the short term, although long-term effects on agriculture may be felt. However, there have been several important areas where new biotechnological techniques have extended those early, but still important, discoveries of baking and brewing.

STARCH PRODUCTS

Starch is a sugar polymer extracted from many plants and seeds.

20

It is an important component of flour and a key energy source in foods like potatoes and bread. Starch itself is an ingredient in many processed and convenience foods (instant soups, gravy powder, etc.) but it can also be used to produce sugar syrups when treated with enzymes which hydrolyse the polymer into its component sugars and others which convert the individual sugar molecules into slightly different ones. In this way 'high fructose corn syrup' (HFCS) is made (from maize starch). HFCS is an important sweetener for soft drinks and other food products and is economically viable to produce in the USA, which grows lots of maize but imports much of its sugar.

FLAVOURS AND COLOURS

Most spices are derived from exotic (and hence expensive) plants, so flavours and also colours are good targets for biotechnology, being more like pharmaceuticals in their market characteristics. There are two main options: microbial production or plant cell culture. Like hybridomas, some plant cells can be grown in flasks and for certain compounds this might offer advantages.

AMINO ACIDS

Amino acids are often produced commercially by fermentation. They are used mainly to supplement animal feeds. Lysine is the most common, although phenylalanine, a component used to make the sweetener Aspartame, has recently become important. Japan is a major manufacturer of amino acids by fermentation. Other organic acids such as citric acid are also made in this way.

SCP AND FOOD SOURCES

One general area for food biotechnology has been developing novel biomass products as foodstuffs. Algae, fungi, bacteria and yeast have been the main organisms used. RHM has developed a fungal 'mycoprotein' which can be textured and flavoured to simulate many meat products. Most of the applications, however, are in animal feed where perhaps the best known example is ICI's 'Pruteen'. Other products have been tried which use bacteria which grow on methanol or other chemical substances, but the economics of these processes have been unfavourable since oil prices are now much higher than cheap feedstuffs like soya beans.

NEW PLANTS

In the long term, biotechnology applied to agriculture will affect food products. At the moment the priorities are plants resistant to insects, drought, salt and the chemical herbicides used to control weeds. Improvements in yield, nutritive value (e.g. corn richer in the essential amino acid lysine) and processing characteristics (e.g. high solids tomatoes for canning) are all currently being developed, and some really novel hybrid plants are being created by cell fusion.

ANIMAL PRODUCTS

Animals used for food production will not remain unaffected by biotechnology. Already, diagnostic tests for pregnancy and oestrus are improving the efficiency of dairy herd management, and soon the use of bovine growth hormone (bGH), produced using genetic engineering, will probably increase milk production by 20–40%. Completely new breeds of animals too are a possibility (a sheep-goat chimera has already been produced), and embryo transfer techniques will make new breeding methods possible.

2.7   Energy and Other Industrial Applications

The simplest way to get energy from biological organisms is to burn wood or other plant matter, but more indirect biotechnology-based methods can provide energy in a cleaner way and allow the use of substrates so wet they cannot be burnt.

METHANE PRODUCTION

Bio-energy is an appealing field leading to better use of 'renewable' resources which are often now wasted. Some applications are closely linked to pollution control (see below) and are based on the ability of many bacteria to live without oxygen ('anaerobically') in pits of sewage sludge or animal feedlot slurry. These organisms often produce swamp gas (methane) which can be burned as an energy source. Similar 'anaerobic digestion' processes can be set up to use starch or sugar-rich wastes from the food industry and even to process household garbage and landfill waste.

ALCOHOL

The fermentation of sugar and starch can also be used to make ethanol – which, of course, can be burned as well as drunk. Alcohol produced in this way can be used as a fuel on its own or mixed with petrol to form 'gasahol'. This mixture is much used in Brazil, a country with little domestic oil production but lots of sugar cane waste. It is also popular in the USA, where maize is used.

POLLUTION CONTROL

Bio-energy as a waste disposal technique has been discussed above, but bacteria also play a vital part in conventional sewage treatment and are increasingly being used to clean up contaminated land or groundwater. Here, unusual strains of microbes which can eat toxic pesticides or chemicals are used. Oil-degrading micro-organisms have also been used to clean up oil spills at sea and on land. Indeed, the first microbe on which a US patent was granted was for this kind of use.

PEST CONTROL

Biological pest control relies on the "big fish eat little fish" principle and involves manipulating the local environment so that the right mix of predators, prey and pathogens are present. Sometimes predators are introduced, but more often bacteria or viruses pathogenic to an insect pest are sprayed on the crop. By far the best known example is *Bacillus thuringiensis*, a bacterium harmless to mammals but which if eaten by the caterpillars or the larvae of moths or butterflies is deadly poison to them. This type of pesticide (made chiefly by Abbott Laboratories and Sandoz) has been on sale for many years and is very effective in controlling the gypsy moth in forests. The toxin is a crystalline protein which can be produced in pure form, but the whole bacteria are cheaper to use and have the advantage of persisting once introduced. Recently, genetic engineers have succeeded in putting genes for toxin production into plants, which could make the plant itself toxic to the caterpillars.

CHEMICALS

Microbes and cells can be used to produce small organic chemicals as well as proteins, and some of the applications already discussed fall into this category (antibiotics, food flavours, etc.). Sometimes useful strains can be isolated, but if genetic engineering is used, whole groups of genes concerned with a complicated metabolic pathway must be manipulated. For this reason, not many chemical processes using biotechnology are commercialised yet, but dyes are one good prospect.

OIL RECOVERY

Oil spill cleanup has been mentioned above. Bacteria can also be used to help extract oil from poor sources. Pumping a cocktail of microbes into an uneconomic oil well can result in more oil being extractable because they produce natural detergents freeing tiny oil droplets from the rock.

PLASTICS AND POLYMERS

Biopolymers are a new area for biotechnology. Many biological materials (e.g. bones, silk, hair) are very well designed and often show astonishing physical properties (the tensile strength of a spider web is quite remarkable, for example). Some of these may be adapted for industrial use. To date the best known example is polyhydroxybutyrate (PHB), a plastic-like substance produced by certain bacteria, being commercially developed by ICI.

GLUES

Some bio-materials also have very good adhesive properties. A sticky substance used by mussels to hold them on to rocks has now been produced using genetic engineering, and may be the basis of a new glue.

MINING

Even metallurgical processes are not exempt from the ever-present microbe. Some bacteria (thiobacilli, mainly) can oxidise iron ores and this can allow low-grade ores to be 'mined' by a solution of micro-organisms. Others can concentrate rare metals such as silver and gold from dilute sources such as sea water.

2.8   Production Techniques

It is beyond the scope of this book to outline what is entailed in the production of biotechnological products in any detail, but a few general principles will be covered here.

FINDING YOUR CELL

Many processes are still dependent on isolating a micro-organism or cell 'line' from nature which does what you want – makes a particular antibiotic, for example. This often involves a process of 'screening', checking through hundreds or thousands of isolates for the one with the desired trait. This is followed by 'cloning', growing a pure culture from one cell. Sometimes the screening involves 'selection', which implies that the growth conditions are altered to favour those organisms with the desired characteristic. This often kills all but a few mutants. Mutation is always occurring, caused by occasional random errors in DNA replication and environmental chemicals and by radiation such as cosmic rays.

These methods are often still used even when genetic engineering is involved. Gene cloning is really just an extension of cell cloning and relies on the same methods of selection to isolate and identify cells which contain recombinant DNA. The basic methods of genetic engineering have already been discussed and it is sufficient here to point out that they are as applicable to animal cells as to bacteria. Bacteria multiply faster, are simpler, and generally have better understood genetics, but animal cells can do some biochemical things micro-organisms can't and hence are better for some applications. Plant cells are difficult to work with because of their very thick cell wall, but this is now being overcome.

GROWING IT UP

Once you have found or made your cell, its best growth conditions must be found. Most micro-organisms can grow either on solid surfaces like agar or in suspension, but some animal cells are more fussy. When large volumes of culture medium are used (100+ litres), aeration is critical but too rough stirring or mixing can damage the cells. Many different designs of fermenter exist, but most comprise a tank either stirred by some sort of paddle or agitated by bubbles from the bottom (an 'air-lift'). Usually the tank is surrounded by a 'jacket', an outer layer containing heated or cooled water to control the temperature in the vessel. Some animal cells are grown on solid supports (such as 'hollow fibres') or are 'encapsulated' in gels. These give 'attachment-dependent' cells something to stick to and increase the available surface area. Smaller-scale culture for laboratory experiments uses smaller fermenters (1–10 litres in volume) or flasks or 'roller bottles' (round bottles agitated by rolling). At this scale, culture is usually on a 'batch' basis where a few cells are inoculated into sterile media and allowed to multiply until their food is gone. Big industrial systems often need to be continuous though, so they will involve a complicated system to monitor the fermentation, removing cells and/or product continuously and adding fresh nutrients.

GETTING THE PRODUCT OUT

Sometimes the desired product are the cells, in which case simply centrifuging or filtering them out will suffice, or they can be used in situ as a 'bioreactor' to carry out a reaction. If the product is a chemical or protein, however, complexities ensue.

Cells are like little porous bags – some substances diffuse out, or are pushed out through the membrane ('secreted'), but others remain inside. The first thing to determine is where the product is: 'intracellular' or 'extracellular'. If it is in the medium, the unnecessary cells may be removed and discarded. If in the cells, the spent medium can be discarded but the cells must be broken open and their contents 'fractionated' to isolate the product. Extra-cellular products usually require concentration and fractionation since many different compounds are secreted and some unused nutrients and debris from dead cells will also be present.

DOWNSTREAM PROCESSING

Extraction and purification of a biological product are referred to as 'downstream processing'. Many techniques are involved, some physical (filtration, drying), others chemical (crystallisation, precipitation by denaturation, chromotography, electrophorsis, etc.). Again, it is outside the scope of this book to explain all these methods, but it is important to realise that every different product, indeed every protein, will require its own purification process and a different one if made in yeast or bacteria, etc. Yields are often quite low and losses high so this stage is of great economic importance. The molecules involved are delicate and must generally be handled within the extremes of temperature, acidity/alkalinity, pressure, etc. found in living organisms, which makes biotechnology production processes very different from tradi-tional chemical engineering. Many different purification methods are often applied sequentially to isolate just the desired product. This highlights one of the disadvantages of using cells as factories – they have to produce a lot of unwanted products as well as the target product. It is as if a car manufacturer wanted to sell only green automatic transmission compacts but had to manufacture all the current product range and then sort them.

3 | Chapter Three

Current Research Trends

3.1 Introduction

The extraordinarily rapid pace of biotechnology research has fuelled the industry's growth and produced a glamour image that sometimes proves hard to live up to: all too frequently, reports of research advances that are years away from testing, much less the marketplace, raise (or sometimes dash) unrealistic expectations about the performance of biotechnology companies.

The sums that have been spent on biotech research are vast – $10s of billions if one includes government as well as private investment in the field – and the time scales from discovery to product are often long – five years to decades. But the prospects for biotechnology's emergence as the 'last major industrial technology of the twentieth century' are sufficiently exciting to continue to capture the imagination (and wallets) of investors, corporations, governments and the public, and to make it one of the most dynamic fields of basic and applied scientific research.

Biotechnology cannot possibly provide 'technofix' answers to all the research goals that have been set for it, but it is providing new discoveries, and in some cases new products, far more rapidly and efficiently than many pundits predicted. For example, recombinant DNA, the basic technology of genetic manipulation, was developed less than 15 years ago as a laboratory technique for basic research on simple, well-defined organisms; it has rapidly evolved into a powerful underpinning technology and is now routinely applied to alter the genetic makeup of a broad range of organisms, including microbes, plants and mammals. Thus research advances in basic biological sciences, particularly cell and molecular biology, have raised hopes for industrial exploitation of the enormous variety of useful products produced by biological organisms, and for 'blue-sky' applications of biological principles and products.

The early 1980s' have brought rash prophecies that biological products and processes would provide cheap and easy solutions to a vast array of industrial problems but these have gone by the wayside, replaced by more realistic evaluations of biotechnology's limited but valuable role in creating novel, high-value products for specialist niches in a spectrum of industrial applications. Research has already produced far more successes than failures, and the industry is now mature enough and secure enough to focus its technological expertise on realistic targets that have a reasonable probability of producing useful products.

This chapter introduces the technologies, applications and markets for biotechnology. These themes are explored in more detail in later chapters, especially Chapter Ten (Market Trends).

3.2 The Economic and Social Context of Biotechnology Research

The development of biotechnology has been driven by advances in basic biomedical and biological research, and whilst biotechnology has emerged as an important industrial sector, it remains closely tied to leading edge research in the biological sciences. The distinctions between basic and applied research have become increasingly blurred as academic and industrial laboratories vie to identify, clone and express the hottest new growth factors or produce the next breakthrough in protein engineering.

University-industry collaborations have mushroomed in the past decade, particularly in the USA where several large corporations have established long-term research programmes within universities, and virtually all companies involved in biotechnology research have forged strong links with academic researchers. Other countries are emulating this trend, and many national and regional governments actively support programmes that encourage links between science and industry. These research partnerships are producing many of biotechnology's most exciting discoveries and reducing the time and costs required to commercialise new technologies.

This trend started in new R & D-based biotechnology companies. Most of these companies have been founded, staffed and sometimes run by research scientists; they have advisory panels of distinguished academic scientists and many of their top researchers hold academic posts and publish their work in academic research journals. The companies quickly produced high-quality, commercially relevant research by attracting outstanding young scientists from all over the world. The success of these R & D-based companies was rapid and well publicised, and goaded large companies into the recognition that biotechnology offered a serious threat to their businesses. Whilst a few large corporations have propelled themselves into the top leagues of biotechnology research, many are still playing catch-up and relying on research, licensing and marketing deals with small companies to provide their first generation of biotechnology products.

This symbiotic relationship between new biotechnology companies that need R & D revenue to augment their long-term product development programmes and large companies that need new technologies and products to augment their own R & D programmes is fuelling much of the research in commercial biotechnology today. Several key factors support this trend:
— Small companies continue to be the most productive source of technological developments and product innovations in many areas of biotechnology research. A few large companies such as Monsanto have developed outstanding research programmes by making a strong, early commitment to biotechnology, but many companies have difficulty recruiting and keeping crea-tive new scientific talent.
— The high cost of R & D has put pressure on both small and large companies to spread the risks of product development. Small companies' resources lie in specialist tools and technologies applicable to a range of research programmes, but most cannot afford to pursue more than a handful of projects on their own; large companies have found that it is often cheaper and more efficient to acquire technology and research-stage products than to develop them in-house.
— The markets for biotechnology are international and highly competitive, and few new biotechnology companies have the resources to market their products aggressively worldwide. Thus many companies support product development by developing strategic alliances with corporate partners in specific geographic markets.
— The uncertainty of government regulatory restrictions on biotechnology has hampered the efforts of small companies to

30

get their products tested and approved. Whilst large companies face the same uncertain regulatory environment, they have the experience and clout to push the approvals process through more effectively and the resources to defend legal challenges to product introduction.

These and other factors have had the effect of focusing most biotechnology research on a small number of high-value markets where novel products carry a large premium and established companies have a tradition of long-term high risk/high reward research programmes. Thus pharmaceutical and health care research has dominated biotechnology and produced most of its early high-profile products. This trend is unlikely to abate: the field is poised to produce its first blockbuster product, Genentech's blood clot dissolver Activase (tissue plasminogen activator), and several other novel biotherapeutics are showing promise in clinical trials. Research in agricultural biotechnology has lagged well behind health care, both in terms of investment and development of new technologies and products. But the pace of agricultural research is accelerating and 1987 has brought the first field trials of genetically engineered plants.

Whilst biotech's big guns are aimed on these two markets, there is a broad spectrum of research targeted at specialist niches in other areas. 'Traditional' biotechnology areas such as enzymes, fermentation and waste treatment are applying new technologies to improve the industrial utility of existing products or processes as well as creating new ones. Research in these fields, however, has traditionally been empirical, and the lack of basic knowledge about the processes has hampered efforts to apply and integrate new technologies. These gaps have stimulated basic research into the physiological and biochemical mechanisms of industrial biotechnology processes, but broad application of new tech-nologies in sectors such as production of speciality chemicals or biological land detoxification will be slow in coming. Finally, government policies and regulations are creating new 'strategic' targets and priorities for biotechnology research: the EEC hopes to reduce its grain mountain by developing biotechnology processes to convert the grain into high-value chemicals, and possibly ethanol; the US government is sponsoring research on microbial recovery of strategic minerals from industrial wastes and low-grade ores; and Japan's MITI has targeted biotechnology as a key industry for the country's economic growth and made government and industrial support for biotechnology research a top priority.

The general expansionary trend in biotechnology research is likely to continue well into the 1990s: not only will biotechnology provide the basis for the next several generations of innovative pharmaceutical and agricultural products, but new discoveries will broaden the scope for applications of biotechnology and improved technologies will reduce the costs and time scales for R & D, making biotechnology products and processes economic alternatives to traditional ones.

3.3   The Key Technologies

Biotechnology draws on the whole spectrum of biological and biomedical sciences: genetics, microbology, biochemistry, cell and molecular biology, zoology, botany, medicine, immunology and other clinical sciences. The long and distinguished history of basic research in these fields has provided profound insights into the hugely varied and complex spectrum of biological processes. But until quite recently, scientists possessed only the crudest tools to probe and manipulate these processes. The explosion of basic biological research in the past 40 years or so has now produced not only a fundamental understanding of biological processes, but also the skills and technologies to modify and adapt them to man's needs.

The 'new era' of biotechnology research began about 15 years ago, when scientific researchers recognised that the sophisticated tools and technologies they were developing to study and manipulate basic biological processes could also be used to create valuable new products. The traditional approach to biotechnology research was empirical: search for an organism that 'naturally' possessed both desirable properties, e.g. the ability to produce an antibiotic or enzyme, and the ability to express these properties under industrial conditions. Researchers would then attempt to adapt and modify the selected organisms to produce their useful products efficiently, but because their tools for accomplishing this were non-specific, the process was very laborious and often unfruitful. Moreover, the range of biological processes they could exploit was limited, since most organisms are adapted to live in the wild rather than a laboratory flask. Advances in three seminal technologies – recombinant DNA, monoclonal antibodies and bioprocessing – have revolutionised this approach, and these technologies now underpin much of the advanced research in biotechnology.

RECOMBINANT DNA

The development of recombinant DNA technology in 1973 is generally regarded as the key discovery that launched new biotechnology because it opened the door to genetic manipulation of a wide spectrum of organisms. In principle the technique is simple: genetic information coding for a useful or desirable trait is sliced out of its host organism, spliced into a vector that makes it easy to manipulate, then transferred to the cells of another organism. Once inside the recipient cells, the information is expressed by the cell, conferring the desirable trait on the recipient organism. In practice, the technique is considerably more complicated, but researchers have devised many clever ways of making the process of genetic manipulation less laborious.

The power of recombinant DNA technology is that it permits researchers specifically to reprogram an organism to produce any desirable or useful biological product. Recombinant DNA technology has enabled researchers genetically to manipulate a wide variety of micro-organisms, animals and plants, reprogramming the cells' genetic information to make them more industrially useful. It has also made it possible specifically to modify the products themselves, giving researchers the power to design new molecules with valuable properties.

MONOCLONAL ANTIBODIES

Research in immunology led to the development of monoclonal antibodies, proteins that recognise and bind to very specific sites on other molecules. When exposed to a foreign molecule, the immune system of higher animals produces an array of antibodies that recognise different bits of the molecule and help remove it from the host organism, constituting a first line of defence against invasion or infection. Many of the antibodies are exquisitely sensitive and specific: they pick out the foreign molecule amongst billions of others and bind to a unique site on it.

Monoclonal antibody technology permitted researchers to exploit the valuable properties of antibodies by removing the antibody-producing cells from the host organism (usually a mouse) and hybridising the cells with tumour cells so they grow in laboratory cell cultures. Because each cell produces only one kind of antibody, the new hybridoma cell cultures provide a virtually unlimited source of a single, highly specific, or monoclonal, antibody. Monoclonal antibodies are now used in a wide range of medical and industrial tests and are being developed as therapeutic drugs and as the recognition element of sensors.

BIOPROCESSING

Bioprocessing is a traditional biotechnology that is being revolutionised by new discoveries in biological sciences. Industrial bioprocesses are systems in which biological organisms or molecules are used to effect chemical or physical changes.

Bioprocesses such as brewing, leavening with baker's yeast and composting are some of man's oldest technologies, and industrial fermentations that produce chemicals such as solvents, vitamins, amino acids and antibiotics have been a mainstay of the chemical industry throughout the twentieth century. But until the advent of new biotechnologies, bioprocesses were economically competitive only for a few products. Rapid advances in researchers' knowledge of how to handle biological cells and molecules under industrial conditions are expanding bioprocessing opportunities considerably. The most important technological developments include:

— Large-scale culture of animal and plant cells that permits production of commercial-scale quantities of valuable cell products such as biopharmaceuticals and monoclonal antibodies.

— Improved fermentations of microbial cells that increase yields and expand the range of organisms available for industrial fermentations.

— Novel, highly specific separation and purification technologies that make it cheaper and easier to isolate pure bioproducts.

— Efficient techniques for immobilising cells and enzymes to create a bioprocessing matrix that permits continuous, high throughput biotransformations.

New developments in other areas of biotechnology research have both created greater demand for bioprocesses that ensure the discoveries become usable products and offered technical advances that expand the range of economic bioprocesses. Whilst bioprocessing research is less glamorous than much of the leading edge research in genetic manipulation or protein engineering, it is essential for the industrial production of any new bioproducts and therefore fundamental to the commercial development of biotechnology.

3.4 Trends in Pharmaceutical and Health Care Research

Given its evolution from basic biological and biomedical research, it is not surprising that many of biotechnology's early research programmes – and successful products – are aimed at health care. The past several years have seen a small but steady flow of biotechnology-based diagnostic and therapeutic products reach the market. But the impact of biotechnology on the pharmaceutical industry has been much more profound: it is changing the whole approach to drug discovery R & D and to the identification and management of disease.

Biotechnology has given researchers the knowledge and tools to understand the processes that underlie many diseases and to design therapies that specifically interfere with them. Basic research in academic laboratories dominates discoveries in·this field, and small biotechnology companies have been quick to exploit its potential. Large drug companies have begun to recognise the enormous potential of biotechnology research and are quickly jumping on the bandwagon labelled rational drug discovery. Thus, while pharmaceutical research will continue to be a high risk/high reward endeavour, the triumph of reason over trial and error has increased the probability of research yielding effective products.

DIAGNOSTICS

Immunodiagnostics are tests based on the ability of antibodies to recognise specific molecules in a sample. They are standard tools in physicians' diagnostic regimens, and improved testing technology is making diagnosis faster, easier and cheaper. Much of the research in this field concentrates on developing tests that can be administered in doctors' offices or even in the patient's home, instead of in hospital laboratories, providing 'instant diagnosis'. Monoclonal antibodies are making the tests more sensitive and specific, and new detection technologies such as fluorescence and enzyme-catalysed colour changes provide fast, accurate answers.

Research to identify tumour-specific markets is fuelling development of new tests for cancer diagnosis, and increased use of immunosuppressive therapies and the spectre of AIDS have increased demand for tests to identify the spectrum of infectious diseases that prey upon immunocompromised individuals. Immunodiagnostics were an early target for biotechnology research because the field offered relatively low barriers to market entry, but the pace of research has slowed as the market matures and new, more exciting and profitable applications emerge. One important new area for monoclonal antibody applications is *in vivo* diagnostic imaging. Monoclonal antibodies attached to powerful signal molecules are injected into the bloodstream and home in on targets such as tumour metastases or damaged tissue, enabling physicians to pinpoint the parts of the body affected by a disease.

DNA probes, a technology described as 'stuck in the lab' just two years ago, have returned to the forefront of diagnostics research. 1986 saw the launch of several DNA probe tests for infectious diseases, including periodontal disease and Legionnaire's disease, but the most exciting research is aimed at detecting genetic abnormalities. Several years ago, genetics researchers developed new techniques for analysing patterns of genetic change in DNA by cutting it into small pieces that reflect genetic polymorphisms (RFLP). They have now identified RFLP associated with a number of heritable diseases, including cystic fibrosis, sickle cell anaemia, haemophilia and Duchenne's muscular dystrophy, and can test for the presence of the genetic defect in a foetus. This type of analysis is also being applied to monitor the success of bone marrow transplants and to identify proprietary strains of micro-organisms or plants.

Researchers are making great strides in identifying the genetic basis of disease, and this information is rapidly being assembled to create prognostic profiles that indicate the presence of both single heritable gene defects and genetic predispositions to complicated disorders such as cardiovascular disease. Whilst genetic screening is a technology fraught with ethical and social overtones, the research is providing powerful new tools to diagnose, and perhaps prevent, genetic diseases.

Research on DNA probes has produced some interesting non-medical applications for the technology. Probe-based genetic analysis techniques can be used to analyse minute samples of blood, semen or tissue for forensic identification and unequivocally to assign paternity – or maternity, in the case of a young Ghanaian who, when challenged by British Immigration authorities, proved that he was the son of a UK resident. In Britain, ICI is establishing a new company devoted to developing these applications, and in the USA Lifecodes has begun offering forensic testing services. Another UK company, Biotechnica Ltd, is developing DNA probes as anti-counterfeiting devices: a unique probe attached to the paper of a valuable document or the canvas of a painting can serve as an invisible but unique identification tag.

THERAPEUTICS

The long time scales for drug development and testing have meant that biotechnology-produced drugs are just beginning to reach the market. Moreover, most of these products are radically different from traditional therapeutics and many have encountered unexpected hurdles in the clinic. 1986 saw the introduction of several significant biotechnology products: two genetically engineered vaccines, interferon alpha and a therapeutic monoclonal antibody, and 1987 is likely to have brought another handful of products to the market.

Much of the focus of drug research has been on artifically engineered versions of natural biological regulators: hormones, immunomodulators, growth factors and key physiological enzymes. Healthy bodies stay that way by maintaining a delicate balance between these regulatory molecules, but diseases disrupt the balance, compromising one or more of the regulatory systems. Biopharmaceuticals attempt to redress the balance by mimicking the body's natural disease-fighting mechanisms. In practice, this rational strategy has been difficult to implement: most of these regulatory molecules have multiple functions and administration of large doses has created unexpected side effects. But whilst no biotechnology-produced drug has yet produced 'miracle cures', many are showing strong therapeutic promise. Table 3.1 lists these drugs, along with one US analyst's ratings of their probability of living up to clinical and market expectations. Top contenders include recombinant Factor VIII, the blood clotting factor used to treat haemophilia, erythropoietin for treatment of anaemia in kidney dialysis patients, tissue plasminogen activator, colony stimulating factors for aiding chemotherapy and bone marrow graft patients, and lung surfactant for treating respiratory distress syndrome in premature infants.

The success of biopharmaceuticals in treating cancer has been mixed: interferons have demonstrated clear positive results in a few rare cancers and recent trials using interleukin-2 (IL-2) in a radical regime termed adoptive immunotherapy have confirmed IL-2's powerful anti-tumour activity. Researchers are also

TABLE 3.1   PROMISING THERAPEUTICS

Product	Application	US market size ($ millions)	Probability
Atrial natriuretic factor	diuretic	45–85	60%
GM-CSF	chemotherapy & bone marrow grafts	125	95%
	AIDS	200	20%
G-CSF	chemotherapy & bone marrow grafts	125	95%
	leukaemia	25	40%
M-CSF (CSF-1)	macrophage stimulation	50–150	50%
Epidermal growth factors	wound healing	150	75%
Erythropoietin	anaemia in kidney dialysis	165	98%
	other anaemias & blood enrichment	160	55%
Factor VIII	haemophilia	155	98%
Follicle stimulating & leutinising hormones	infertility	65	95%
Human growth hormone	short stature	300	80%
Interferon alpha	cancer	80	85%
		>150	50%
	infectious disease	100	70%
Interferon beta	cancer & infectious disease	20–30	60%
Interferon gamma	cancer (monotherapy)	45	60%
	arthritis	25	40%
Interleukin-2	cancer	50–75	80%
		300–500	50%
	infectious disease	35–70	40%
Lipocortin	anti-inflammatory	50	35%
Lung surfactant	infant respiratory distress syndrome	55	95%
Monoclonal antibodies	cancer therapy	100	80%
		300	50%
	antibacterial therapy	180	80%
	antibacterial prophylaxis	300	45%
Superoxide dismutase	heart attack	250–350	60%
	other therapies	160	40%
Tissue plasminogen activator	blood clots	400	95%
		790	70%
Tumour necrosis factor	cancer (monotherapy)	50	60%

Source: Robert Kupor, Cable Howse & Ragen

beginning to achieve positive results with combination therapies, where several drugs administered together seem to act synergistically, but working out the ideal combination of drugs could take many years. The largest barrier to developing good cancer therapies is our limited understanding of the nature of the disease, and whilst the last few years have provided a wealth of knowledge about oncogenes and some of the molecular mechanisms that underlie the transformation of a normal cell to a malignant one, this has yet to contribute much insight into how to treat the disease that results.

The mixed results of clinical trials with this first generation of biotechnology therapeutics have stimulated the search for ways to alter or modify the products to improve their performance – and to create unique, patentable compounds. Because most biotherapeutics are large proteins, they offer problems of administration, instability in the bloodstream and antigenicity due to microheterogeneity. Researchers are developing a variety of approaches to solving these problems. Scientists at Cetus have created mutants of naturally-occurring proteins that appear to have improved stability, but most researchers think this approach has limited value. A more promising approach is to identify fragments or different versions of the large molecules that possess the key pharmacological activity but induce fewer side effects. For example, researchers have identified several alternative forms of TPA, and claim that some are more specific than the current product.

But neither proteins nor small peptides are really very good drugs: an ideal drug can be administered orally, which means it cannot be subject to proteolytic breakdown and is small enough to be absorbed effectively. The need to find these kind of molecules has led researchers back to 'traditional' sources of drug discovery, but with a new approach. This new approach is exemplified by a recent report from researchers at Merck, who identified an effective inhibitor of cholecystokinin (CCK), an intestinal hormone that stimulates bile production and affects several neurological functions. Knowing that CCK acts by binding to cell surface receptors, the researchers designed sensitive assays for receptor binding, then screened microbial broths for compounds that blocked this binding. They identified a small, heterocyclic molecule, asperlicin, then characterised its chemical structure and began modifying it chemically. They eventually developed a product that was much more potent than the original asperlicin but retained its CCK-antagonising properties. Whilst the clinical effectiveness of the compound has yet to be demonstrated, it represents a landmark in rational drug design: identification of a molecule with specific therapeutic potential via a targeted screening programme.

TARGETED DRUGS

Targeted drug delivery has long been a goal of pharmaceutical research, and biotechnology is starting to provide tools for accomplishing just that. The exquisite specificity of monoclonal antibodies makes them an ideal vehicle for delivering drugs to specific cells. For example, most chemotherapeutic cancer drugs work by killing cells, but none is specific enough to kill only tumour cells. If such a drug is attached to a monoclonal antibody that recognises only tumour cells, it will be far more effective and produce fewer side effects because most of the drug will end up at the tumour.

The need to deliver biotherapeutics effectively has stimulated research on other drug delivery systems. Researchers at California Biotechnology have developed a way to deliver insulin via nasal inhalation: the hormone is suspended with another chemical that facilitates its absorption into the blood. Other researchers are developing polymers that entrap drugs in a matrix and slowly release them into the bloodstream. Although still in its infancy, drug delivery research is likely to become far more important as biotechnology produces novel drugs that require new and carefully controlled administration.

VACCINES

Vaccine development is a field where biotechnology will have enormous long-term impact. Vaccines represent one of the most powerful tools we have for controlling infectious diseases, particularly in the lesser developed regions of the world where health care is at best rudimentary. But the high cost of developing and producing safe, effective vaccines, and fear of liability problems and consumer resistance, made all but a handful of pharmaceutical companies reluctant to commit to new vaccine development. This attitude is beginning to change: researchers have discovered powerful new biotechnological approaches to vaccine development that increase the safety and efficacy of vaccines whilst reducing the cost of producing them. Researchers have already used genetic manipulation to produce a handful of new human and animal vaccines, and are making significant progress on vaccines against parasitic diseases such as malaria and schistosomiasis. Moreover, the spread of AIDS has increased public awareness of the threat of untreatable diseases, and the potential role of vaccines in controlling them.

Three strategies for novel vaccine development have shown real promise. The first is to produce immunising subunits via recombinant DNA; this approach was used for the hepatitis B vaccines very successfully, and is now being applied to a range of viral, bacterial and parasitic diseases. The second is to modify the smallpox vaccine by introducing foreign genes that direct it to produce immunising proteins from other viruses; this approach produced the first potential AIDS vaccine, which is undergoing preliminary tests in Zaire, and in the long term could produce 'polyvalent' vaccines that immunise against a spectrum of diseases with a single injection. A more complicated approach involves the production of anti-idiotypic antibodies, which mimic the immunising elements of pathogens, and offer the possibility of immunisation without direct exposure to any part of the pathogen. This approach holds particular promise for immunisation against complicated and dangerous viruses such as those that cause AIDS or human cancers.

3.5 Trends in Agricultural and Veterinary Research

Agricultural biotechnology is producing some of the most exciting research in biology: researchers have made significant strides in developing tools and strategies to improve and protect commercially important crops and livestock. But much agricultural biotechnology research is stuck in the laboratory, unable to enter field trials because of restrictions on the environmental release of genetically engineered organisms. US researchers are seriously concerned that the lack of a clear regulatory policy will keep their experiments indoors for years. This fear is not unfounded: Agracetus has waited three years to obtain approval for field tests to determine how well they can follow tobacco manipulated to contain a simple marker gene.

The regulatory situation in Europe is more fluid: the UK last year approved phase one field trials to study the survivability of a genetically marked insect virus, and approval is pending for field tests of several new organisms, including genetically altered potato plants. Powerful environmental lobbies in several European countries are organising to block release of genetically engineered organisms, and if they can successfully convert public concern over the environmental impact of such trials into a political issue, the experiments could be shelved for years.

PLANT IMPROVEMENTS

One area of biotechnology research that has not been hindered by regulatory constraints is non-genetic manipulation of plants in cell cultures. Many economically important plants can be grown as laboratory cell cultures, which are relatively easy to manipulate, then induced to develop into viable plants. Researchers have exploited this useful property of plant cells to develop a variety of techniques for crop improvement. One of the most powerful techniques is somaclonal variation, which takes advantage of the naturally-occurring genetic variability of plants regenerated from tissue culture cells. Researchers have developed ways to control and manipulate this variability, e.g. by changing the culture medium and using cells derived from very young or very old plants, to create new varieties with improved characteristics. The scope of improvement using this technology is limited by the natural genetic makeup of the plants, but is has already produced commercially valuable varieties of vegetables, and could be used to improve some cereals and other commodity crops.

Researchers have also developed ways to produce artificial hybrids of plant cells grown in culture: protoplasts, individual plant cells stripped of their cell walls, can be artificially fused in the laboratory to produce cybrids, new cells that incorporate characteristics from both parent cells. This technology is particularly useful for transferring valuable traits (disease resistance, herbicide resistance, male sterility) carried by the genes of cytoplasmic organelles of many species, because the transfer does not necessarily involve genetic rearrangements.

Applications of this technology continue to be limited by the difficulties in regenerating plants from protoplasts, but research in this area is progressing. For example, last year two Japanese groups reported the successful regeneration of rice plants from protoplasts, raising the possibility of controlled manipulation of this staple food of half the world's population. Plant protoplasts also offer a useful target for genetic manipulation techniques: unlike normal plant cells, protoplasts can be induced to take up foreign DNA. Whilst research on genetic manipulation of protoplasts is still in its infancy, the technology holds great promise for genetic engineering of cereal crops, many of which are refractory to current gene transfer systems.

New varieties produced from plant cell cultures are still generally propagated via conventional means, i.e. grown into mature plants that produce seed for future generations. But biotechnology researchers want to bypass this time-consuming process by producing artifical seeds, tiny plant embryos encapsulated in a protective gel. Thus researchers are concentrating on developing economic large-scale systems to produce plant embryos and new techniques for protecting the embryos while they regenerate in the field or nursery. This technology is being applied to plants produced by clonal propagation, when tissue from a single parent is cultured and used to propagate genetically identical embryos, or 'difficult' hybrids that do not readily produce marketable quantities of seed.

PLANT GENETICS

Regulatory restrictions are not the only barrier to application of recombinant DNA technology to agriculture: development of genetic manipulation systems has been hampered because so little is known about the molecular and cell biology of plants. Thus, much of the research in this field has concentrated on understanding the mechanisms of gene expression in plants and developing the systems and tools to effect gene transfer. By far the best developed system for genetic manipulation of plants is the Ti plasmid derived from a soil bacterium that infects wounded plants. Researchers have spent several years reconstructing this plasmid, which has the ability to reprogram a plant cell to produce its gene products, and make it an efficient vector for genetic manipulation. Much of the early work used technically useful but commercially uninteresting genes; researchers are now turning their hands to applying this system to genes with more useful properties.

Researchers have identified several genes that confer resistance to crop diseases or pests. The bacterium *Bacillus thuringiensis* (BT) is already used to control numerous insect pests: it produces a toxin that poisons insects but does not affect other animals. The gene coding for this toxin has been isolated and transferred to several plant species. The toxin is produced by cells throughout the plant and early greenhouse work indicates the genetically engineered plants are resistant to economically important insects such as tobacco horn worm and brassica pest. Similar genetic engineering techniques have been used to produce plants resistant to infection by tobacco mosaic virus (TMV). Earlier work showed that exposing plants to mild strains of certain plant viruses confers cross-protection against subsequent infection by more virulent strains. Researchers attempted to mimic this phenomenon by transferring the gene for a TMV protein into tobacco cells. The plant cells produce large amounts of the protein, which confers limited protection from infection by the virus.

These early successes demonstrate that the possibility of engineering genetic resistance to crop diseases and pests is likely to become a reality. Moreover, recent research has extended considerably the range of crops amenable to genetic manipulation using the Ti plasmid. Ordinarily, only broad-leaved plants (dicots) are susceptible to genetic transfer using this system; most of the world's main food crops (cereals, maize and rice) are monocots and resistant to the genetic invasion of this vector. However, the dividing cells that produce pollen in these plants are susceptible to injection of foreign DNA, and researchers exploited this susceptibility successfully to transfer a gene for antibiotic resistance into rye plants. Whilst the broad applicability of this technique has yet to be shown, this discovery offers the tantalising possibility of a simple genetic transformation technique for cereals.

Improving the quality of crops with genetic engineering is a longer-term goal: producing a grain with improved protein content or altering the chemical composition of oil from seed crops requires an understanding of complex metabolic pathways. But research in these areas is being driven by development of new technologies for understanding and manipulating plant metabolism, and the rapid progress of this research suggests that it may provide the next major breakthrough in agricultural biotechnology.

PEST CONTROL

Biotechnology is also producing valuable new tools and products for biological control of agricultural pests and diseases in integrated pest management strategies. The economic and environmental consequences of increasing resistance to chemical pesticides have created demand for more rational use of these chemicals; the controlled use of biological predators and pathogens in conjunction with selective application of chemicals is proving a successful alternative strategy. Scientists are using genetic engineering to improve the safety and efficacy of viruses and bacteria that prey on insect pests, in the hope of using these organisms to control pests in the wild. Biotechnology is also being used to identify and produce natural substances that help plants resist infection and infestation, and to protect them from environmental assaults such as frost and drought.

LIVESTOCK

Animal health research in biotechnology is being driven by the need to improve the economics of animal production whilst satisfying consumer demand for reductions in the use of animal growth-promoting compounds such as hormones and antibiotics. Biotechnology is already producing valuable animal health products: vaccines against livestock diseases were some of the first products of genetic engineering, and animal equivalents of several human biopharmaceuticals are showing promise in treatment of diseases such as shipping fever.

Probiotics, mixtures of selected micro-organisms that improve feed conversion rates, provide an alternative to extensive use of antibiotics in animal feed, and researchers are developing new strains and combinations that are more effective. The recent EEC ban on steroid hormones for growth promotion has stimulated research into species-specific peptide growth promotants as well as synthetic molecules that selectively affect metabolic processes. Many of these compounds produce an added attraction: the animals, and the meat they produce, are leaner.

One set of compounds of particular interest are the somatotrophins, naturally produced peptide growth hormones. Bovine somatotrophin can significantly increase milk yields from dairy cows and is currently being tested in the USA and UK. Similar compounds derived from other species are showing promise for producing cheaper, leaner meat.

Ultimately the application of biotechnology to agriculture will depend on consumers: technology and intensive farming have produced extraordinary levels of nutrition and food availability in the developed world, but these changes have radically altered the nature and economics of agriculture, and created a production capacity that far outstrips demand for many agricultural products in the rich North. It is government policies, not new technologies that have created vast surpluses, but public reaction to butter mountains and grain dumping has created a poor environment for introduction of new technology in agriculture and hindered the transfer of agricultural research from the laboratory to the farm.

3.6 Industrial Applications

Enzymes are generally far more efficient and selective than chemical catalysts, but only a few of the tens of thousands of useful enzymes have found industrial applications. This is because most enzymes are very fastidious: they work only in a narrow range of environmental conditions. Moreover, enzymes are large, fragile molecules that tend to denature when confronted with the abuses of heat, solvents and extreme conditions that characterise most industrial processes.

Researchers are now making progress in engineering enzymes that can resist these abuses: they alter the genetic information coding for an enzyme in order to introduce new chemical properties into the molecule, e.g. new chemical bonds that stabilise its structure. Because proteins are large, complicated molecules and protein engineering research is still in its infancy, success in enzyme modification has been limited. But this is an exciting field with great potential, and is likely to produce significant discoveries in the next 3–5 years. Enzyme researchers are also looking for new catalysts engineered by nature: microbes that live in extreme environments such as thermal springs must have enzymes that survive these high temperatures, and these enzymes may find applications in industry.

Biosensors are another emerging area of biotechnology research. Biosensors exploit the sensitivity and selectivity of biological molecules to detect specific molecules; the biological 'recognition' molecule is coupled to an electronic sensor, which registers the recognition event. A few biosensors are already used widely in medical and research applications, but new technologies for coupling proteins such as monoclonal antibodies and enzymes to electronic sensors are expanding the range of biosensor applications and reducing the costs of developing and producing the products.

The food industry traditionally has been the largest user of biotechnology, but its conservatism and relatively low investment in research have slowed the application of new technologies. Nevertheless, biotechnology research is identifying new, natural flavours, fragrances and colours, particularly new sweeteners, producing new rapid tests for food contaminants or diluents and creating new processes to convert food wastes into high-value products.

Biotechnology is also being applied to solve environmental problems: researchers are identifying and improving organisms that degrade toxic chemicals, selectively accumulate toxic metals and provide improved water and waste treatment systems.

CONCLUSION

Biotechnology research has proven most successful when it is applied to biological problems amenable to rational analysis, but the field is still dynamic and flexible enough to respond to serendipitous opportunities to provide biological solutions to industrial problems. The movement of biotechnology from the laboratory to the marketplace will accelerate, not slacken the pace of research, for each discovery elucidates the enormous potential of biological processes.

Part Two
The Major Industries: USA, Japan, Europe

4 Chapter Four
The Biotechnology Industry in the USA

4.1   Introduction

Biotechnology is not considered as a separate industry by the US Department of Commerce in collecting statistics about industry structure and output. Rather it is viewed as an array of multidisciplinary technologies that can be applied by a number of different industries, including (but not limited to) pharmaceuticals, chemicals, agriculture and agricultural processing. Thus, traditional industry statistics available from US government sources are of little assistance in either identifying participants or determining the scope and size of this emerging sector.

Further complicating the task of developing a statistical profile of the biotechnology sector, there is some degree of disagreement concerning the scope of activities and technologies that are approximately included in the biotechnology industry. At one extreme are those analysts who consider any use of enzymes, plant and animal breeding techniques and other biological processes to be an example of 'biotechnology'.

At the other extreme, the Department of Commerce, in its attempts to profile the industry, has used a narrow definition of biotechnology that will apply in this analysis. It confines its analyses of the biotechnology sector to firms and/or products that utilise molecular and cellular manipulation (including recombinant DNA and RNA), enzyme technology (not including traditional fermentation techniques, such as those utilised by the liquor manufacturing industry), microbial technology and bioprocess engineering. Thus, utilisation of the advances in biological techniques that have been made over the past two decades determines whether a firm or product is treated as part of the biotechnology sector.

The biotechnology industry in the USA is acknowledged to be the world's leader in terms of R & D, activity and production for the marketplace. The US biotech industry also leads in patent generation, both for products and processes. For example, among the biotechnology patents issued in the USA in 1985, US corporations and research organisations accounted for 59 per cent. The next largest proportion went to Japanese organisations/corporations (17 per cent) and West German entities (7 per cent) (see Table 4.1). The basis for the US lead has been a well-funded government research programme (mainly basic research) and a strong, entrepreneurial effort by researchers (many from the public sector and/or universities) to commercialise the advances in techniques and knowledge that they have pioneered.

TABLE 4.1   BIOTECHNOLOGY PATENTS ISSUED IN THE UNITED STATES BY COUNTRY OF ORIGIN

Country	Percent of all US patents
United States	59
Granted to	
US corporations	45
US universities	8
Other US organisations	6
Japan	17
West Germany	7
France	3
United Kingdom	3
Canada	2
All other countries	9

Note: A total of 1078 biotechnology patents was issued in 1985
Source: US Patent Office

4.2   Structure of the Biotechnology Industry

Although precise figures are unavailable, more than 400 American firms are actively engaged in biotechnology research and production. (In addition, approximately 200 firms are involved in the supply of production equipment to this sector.) About seventy-five percent of these firms are entrepreneurial, 'startup' organisations that were established specifically for research/

production of biotech products and production technologies. The remaining firms are major producers of chemicals, pharmaceuticals, agricultural products, foods and other products.

Through the end of 1986, about 75 biotech startups had raised capital through public stock offerings. These firms include the largest biotechnology companies in terms of revenues. However, even the largest publicly-held startups are not large by US corporate standards. In 1985, only three had revenues in excess of $50 million, with the largest startup – Genentech – recording revenues of just under $90 million. Table 4.2 presents information on 1985 revenues of selected biotech startups, including the largest publicly-held companies. Only eight recorded revenues of $20 million or more.

**TABLE 4.2   FINANCIAL RESULTS FOR SELECTED
BIOTECHNOLOGY STARTUP FIRMS, 1985
(all figures in million $)**

Firm	Total revenues	R & D expenditures	Profit (loss)
Amgen[1]	23.4	17.4	0.5
Applied Biosystems[1]	59.2	8.0	10.4
Biogen	21.4	31.0	(19.1)
Bio Response	2.9	4.7	(3.8)
California Biotechnology	9.6	8.8	(0.5)
Centocor	22.4	9.2	3.5
Cetus	50.0	32.3	1.1
Chiron[1]	7.1	9.7	(4.2)
Collaborative Research	9.5	3.0	(6.9)
Damon Biotech	3.7	5.4	(8.1)
Diagnostic Products	22.7	3.3	3.9
DNA Plant Technology	4.5	3.8	(0.4)
Genentech	89.6	64.9	5.6
Genetic Systems[2]	9.0	10.2	(3.9)
Genex	16.2	7.7	(15.9)
Hybritech[3]	30.8	13.7	0.4
Molecular Genetics	8.3	5.3	(2.5)
Monoclonal Antibodies	9.5	1.6	(0.1)
Ribi ImmunoChem	1.0	0.5	(0.5)

Notes:
[1] Results for 1986
[2] Results for 1984. Genetic Systems was acquired in late 1985 by Bristol-Myers for $294 million.
[3] Results for 1984. Hybritech was acquired by Eli Lilly for nearly $400 million.
Source: Annual Reports

Biotechnology revenues for established (i.e. non-startup) companies are not available by firm, since these organisations seldom publish sales figures by product line or production methods (i.e. biotechnology versus all other). However, the US Department of Commerce has compiled revenue information for this sector based on primary and secondary research, using the Department's narrow definition for this sector (see Introduction). These figures are reported in Table 4.3. As the table shows, biotech product revenues have grown extremely rapidly since 1984, increasing by nearly 570 per cent to more than $400 million in 1986. These product sales figures do not include revenues derived from R & D contracts and sales of production and research and development equipment to the biotechnology sector. Adding revenues from these two sources more than doubles the industry's revenue base in 1986 to nearly $1 billion (up from less than $450 million in 1984).

TABLE 4.3   ESTIMATED REVENUES FOR THE BIOTECHNOLOGY INDUSTRY FOR 1984–1986 (in millions of $)

Revenue source	1984	1985	1986
Product sales	60	220	400
Contract research revenues	200	250	307
Equipment/instruments	175	210	265
Total revenues	435	680	972

Source: US Department of Commerce

One of the major factors involved in the rapid increase in product sales has been the approval by US government agencies of several biotechnology products, such as diagnostic tests and therapeutic drugs, for full-scale commercial production. To a great extent, the rapid increase in sales has been facilitated by the tendency of biotech startups to transfer production rights for biotechnology products to larger companies through licensing, joint venture or other contractual arrangements. In particular, research contracts have proven to be a major source of funding for biotech startups. As Table 4.4 shows, most major biotech startups in 1985 derived more revenues from contract research than product sales. In fact, in 1986, for the first time, revenues from biotechnology product sales exceeded those from contract revenues (see Table 4.3).

**TABLE 4.4   REVENUE SOURCES FOR SELECTED
BIOTECHNOLOGY STARTUP FIRMS, 1985**

Firm	Percent of total revenues		
	R & D contracts	Investment income	Product sales & royalties
Amgen[1]	86.7	9.7	3.6
Applied Biosystems[1]	1.2	9.0	89.8
Biogen	70.3	29.7	0.0
California Biotechnology	78.8	21.2	0.0
Centocor	57.9	10.7	31.4
Cetus	74.3	20.1	5.6
Chiron[1]	78.5	12.3	9.2
Collaborative Research	27.1	19.1	53.8
Damon Biotech	20.3	56.5	23.2
Diagnostic Products	0.0	0.0	100.0
DNA Plant Technology	69.1	30.9	0.0
Genentech	85.3	8.9	5.8
Genetic Systems[2]	60.3	39.7	0.0
Genex	17.4	(4.2)[4]	86.8
Hybritech[3]	48.1	8.7	43.2
Molecular Genetics	57.9	25.1	17.0
Monoclonal Antibodies	18.6	1.1	80.3
Ribi ImmunoChem	2.8	60.8	36.4

Notes:
[1] Results for 1986
[2] Results for 1984. Genetic Systems was acquired in late 1985 by
Bristol-Myers for $294 million.
[3] Results for 1984. Hybritech was acquired in late 1985 by Eli Lilly for
more than $300 million.
[4] Indicates a loss for investment income
Source: Annual Reports

Table 4.3 also gives data on biotechnology instruments and
production equipment. While revenues for these products have
not increased as fast as those for product sales, instruments and
production equipment still accounted for more than one-fourth
of total biotechnology revenues in 1986.

One final observation concerning the activities of large, established manufacturers in biotechnology is important. Although the
small, entrepreneurial firms gain most of the publicity for their
advances in the field, most of the companies granted multiple
patents in 1985 were large, diversified pharmaceutical companies.

In fact, of the 32 firms receiving five or more biotechnology patents in that year, only four were small, entrepreneurial startups (Eli Lilly led with 28 patents, followed by Miles Laboratories with 19).

4.3 Major Market Sectors

Research and development in the biotechnology field in the US has been focused on three major categories of potential applications — pharmaceuticals, agricultural products, and chemicals — and R & D and commercialisation efforts are most advanced in these fields. Table 4.5 presents a compilation from the US Department of Commerce on potential applications for biotechnology research being pursued by US companies.

TABLE 4.5 POTENTIAL INDUSTRIAL APPLICATIONS OF BIOTECHNOLOGY

Industry	Products	Potential use
PHARMA-CEUTICALS	Hormones (e.g. insulin, endorphins, epidermal growth factor)	Treatment of a wide range of human disorders
	Interferons	Treatment of cancer & infections
	Tumour necrosis factor	Treatment of cancer
	Interleukin-2	Treatment of cancer
	Colony stimulating factor	Treatment of immune suppression
	Plasma derivatives	Blood expanders, treatment of haemophilia & other disorders
	Urokinase	Dissolve blood clots
	Monoclonal antibodies	Diagnostics, in vivo imaging, and drug delivery
	Antibiotics	Treatment of infectious disease
	Vaccines	Immunisation—hepatitis, herpes, influenza, AIDS, etc.

TABLE 4.5 POTENTIAL INDUSTRIAL APPLICATIONS OF BIOTECHNOLOGY *continued*

Industry	Products	Potential use
AGRI-CULTURE	Plant cultivation	Nitrogen fixation capability, tolerance for high salinity, low moisture, resistance to disease & pests, increased nutritional qualities
	Micro-organisms	Natural pesticides, nitrogen fixation
	Animal husbandry	Vaccines, diagnostics, therapeutics, growth hormones
	Aquaculture	Vaccines, growth hormones
CHEMICALS	Amino acids	Feed additives/supplements, artificial sweetener
	Aroma chemicals	Flavours, fragrances
	Bipolymers	Oil cleaning/recovery, biodegradable plastics, food processing/additives
	Biomass conversion	Production of bulk and intermediate chemicals
	Enzymes	Production of sugars, phenols, glycols, detergents, cheese
	Oils, fatty acids/alcohols	Production of industrial chemicals
	Single cell protein	Production of protein supplements
	Vitamins	Nutritional supplements
OTHER SECTORS	Biosensors	Medical diagnostics, monitor industrial processes, air/water quality, detect chemical/biological warfare agent
	Biochips	Electronic applications
	Chemical degrading organisms	Toxic waste and pollution control
	Energy/biomass conversion	Production of fuels
	Metal leaching	Metal recovery
	Mineral concentrators	Mineral recovery
	Fibres	Production of raw fibres for textile manufacture

Source: US Department of Commerce

At this time, commercialisation efforts for biotechnology in the USA are most advanced for therapeutic pharmaceuticals, particularly for the treatment of cancer, development of hormones for treatment of hormonal imbalance-induced conditions in humans and the development of diagnostic tests. In addition, the development of vaccines for humans and animals has progressed rapidly in the last few years, so that vaccines based on recombinant organisms have recently been introduced into the commercial market. Finally, the development of genetically altered plants is near the commercialisation stage.

Among the biotechnology products approved for commercial distribution and sale during the past few years are:
- numerous monoclonal antibody-based diagnostic tests for humans (e.g. in-home pregnancy tests and tests to identify the presence of AIDS, antibodies in donated blood) and animals;
- vaccines for use in humans (for hepatitis B) and animals (e.g. swine herpes);
- monoclonal antibody treatment of diseases in humans (e.g. to prevent rejection of transplanted kidneys) and animals; and
- synthetic production of hormones (e.g. insulin and human growth hormone) for treatment of hormone-related disorders.

At this time, most of the product revenues recorded for fully commercialised biotechnology products (excluding biotechnology instruments, laboratory supplies and production equipment) are derived from products related to the diagnosis or treatment of human diseases and conditions. Most of the remaining sales revenues are derived from either food processing applications (mainly products used in the production of aspartame, a non-caloric sugar substitute) or other agribusiness applications (mainly products for the testing, treatment or prevention of diseases in animals).

In terms of the technologies and/or production techniques used to manufacture these commercial products, monoclonal antibody techniques dominate the market. Approximately 75 per cent of all commercial sales in 1985 for biotechnology products were for products produced using monoclonal antibody techniques. Most other sales were for products using recombinant DNA techniques.

Because monoclonal antibodies are used mainly in diagnostic tests (many of which are conducted outside the human body in lab instruments), approval for these products has been far simpler and faster than for many recombinant DNA products, which often require either release into the environment of genetically-altered organisms or injection into the human body. Thus, to date monoclonal products have penetrated the commercial marketplace in far greater numbers than recombinant DNA products.

TABLE 4.6   STARTUP BIOTECHNOLOGY FIRMS, BY TECHNOLOGY EMPHASIS

Firm	Market valuation[1]
A. rDNA	
Advanced Genetic Sciences	54
Amgen	139
Biogen	281
Biotech Research	49
Biotechnica International	41
Biotechnology General	39
California Biotechnology	93
Cetus	673
Chiron	89
Collaborative Research	58
Cooper Biomedical	32
Enzo Biochem	131
Genentech	1187
Genex	28
Integrated Genetics	43
Molecular Genetics	70
B. Antibody production	
Bio-Response	66
Cambridge BioScience	21
Centocor	194
Damon Biotech	174
Genetic Systems	198
Hybritech	316
Monoclonal Antibodies	43
Summa Medical	38
C. Other emphasis	
Applied Biosystems	339
Genetic Engineering	3
Immunex	83
Interferon Sciences	23
Ribi ImmunoChem	26
Vega Biotechnologies	5

Note: [1] Figures in million $ as of December 1985
Source: Various industry estimates

4.4 Corporate Profiles

Participants in the USA biotechnology industry can generally be classified into one of two categories:
– large, established corporations; or
– small, entrepreneurial firms.

In addition, each of these categories can be further subdivided. Among established, major corporations, participants are usually either:
– drug/chemical companies with products that compete directly with new or potential biotechnology applications; or
– companies with strong materials sciences orientations; or
– agricultural products/food processing companies.

Small, entrepreneurial participants may either be publicly held or privately owned. Most of these firms are developing products using bioengineering techniques, although a few specialise in developing instruments and equipment for this sector.

LARGE ESTABLISHED CORPORATIONS

The large, established corporations which have entered the biotechnology field have generally pursued one or more of several routes of entry:
– joint ventures with small, entrepreneurial firms;
– equity investments in small, entrepreneurial firms;
– takeovers of small firms;
– support of basic and applied research at universities or private laboratories (including small biotech startups);
– support of research at company-owned facilities; and
– establishment of small, 'intrapreneurial' product development groups within the corporation.
Examples of each type of activity are numerous.

For example, Schering-Plough, among other recent investors in biotechnology, has been licensed to produce alpha interferon for treatment of certain cancers and spent nearly $30 million to buy DNAX, a laboratory that conducts research in cellular and molecular biology. American Cyanamid has agreed to pay Celltech Ltd (UK) $7.5 million for an initial two-year research project to develop genetically engineered monoclonal antibodies aimed at identifying and killing tumorous cells. American Cyanamid has sponsored numerous other research contracts with various biotech startups as well. Abbott Laboratories, among its various biotech investments, recently invested more than $20 million in new biotech-related facilities (included four fermentors) to expand product capabilities for genetically engineered products.

Other companies have pursued the buyout route. In one of the most expensive deals to date, Eli Lilly in 1986 purchased Hybritech for nearly $400 million, while many competitors have purchased minority stakes in biotech companies. For example, in 1985 Syntex bought 18 per cent of Genetic Systems for about $40 million (Genetic Systems was subsequently purchased by Bristol-Myers).

Celanese, meanwhile, announced in September of 1986 another approach. It helped establish a new biotechnology company, Celgene, by contributing its six-year, $20 million speciality biotechnology programme (including patents, equipment and products) in exchange for a 47 per cent interest in the new company.

These moves by large chemical, drug and other companies point to a rapidly growing trend by many major US corporations to diversify their production capabilities by utilising biotech production processes and technologies. Two of the most aggressive diversification efforts have been undertaken at Eastman Kodak and Monsanto. Eastman Kodak, with annual sales in 1985 of about $10.6 billion, has not fared well recently in its core photographic equipment market. It had to withdraw from the instant photography market after losing a patent infringement suit to Polaroid and has been losing share in the film market to Fuji. Thus, it was somewhat of a surprise to industry analysts when Kodak announced plans to become a force in the biotechnology market in 1985. Using its strength in materials technology and chemicals manufacturing and its in-house Life Sciences Division (created in 1984) as a base, Eastman Kodak has committed more than $200 million to the biotech field. Among its ventures to date:

- a $45 million joint venture with ICN Pharmaceuticals to develop new drugs.
- purchase of 16 per cent of Cytogen for $15 million.
- investment of $6 million in a joint research project with Cytogen to apply monoclonal antibodies to cancer therapy and diagnostic imaging.
- purchase for $6.6 million of 10 per cent of Elan, which is developing new drug delivery technology.

Kodak has also entered into joint agreements with several biotech firms, including Amgen (for developing biosynthetic processes for production of speciality chemicals), Molecular Genetics (to develop a monoclonal antibody for treatment of bovine diseases), Immunex (to commercialise lymphokines) and Cetus (to develop in-vitro diagnostics). Kodak hopes to register $1 billion in revenues from biotechnology and related products by 1993.

Monsanto is one of several of the traditional chemicals manufacturers (Du Pont is another) that has been attempting to diversify away from commodity chemicals to higher value-added activities, including biotechnology. As part of its biotech strategy, Monsanto has doubled R & D spending in the last three years and will spend about $100 million on biotechnology-related research alone in 1986. In addition, the company recently completed a $150 million life sciences laboratory. Among its major biotech ventures are the purchase (for $2.7 billion) of G D Searle (a pharmaceutical maker and early biotechnology leader) and a $62 million collaborative research effort with Washington University to develop biotech applications for immunology, cardiovascular and central nervous system research. Monsanto has also made equity investments in several biotech startups, including Biogen, Genex, Genentech and Collagen. Besides its research on products for treatment and diagnosis of human diseases, Monsanto is very active in developing products with agricultural applications, such as bovine growth hormones, genetically engineered plants which are herbicide and virus resistant and microbial pesticides.

Tables 4.7–4.9 present information on selected large companies that are active in the biotechnology field.

SMALL ENTREPRENEURIAL COMPANIES ('STARTUPS')

Among all the biotech startups that have formed over the past 15 years, Cetus and Genentech dominate the industry in terms of sales performance and market valuation. It has been estimated that these two companies account for up to 40–50 per cent of the market valuation of all publicly-held biotech companies.

Genentech was formed in 1976 by a venture capitalist, Robert Swanson, and University of California scientist Herbert Boyer. In 1977, under Boyer's direction, their laboratories produced the first genetically engineered human protein, and in 1978 Genentech, under a contract with Eli Lilly, cloned human insulin. By

TABLE 4.7 REVENUES, PROFITS AND RESEARCH AND DEVELOPMENT EXPENDITURES FOR SELECTED DRUG COMPANIES PARTICIPATING IN THE BIOTECHNOLOGY MARKET, 1985
(in millions of $)

Company	Revenues	Profits	R & D exp.
Abbott Laboratories	3360.3	465.3	240.6
American Home Products	4684.7	717.1	217.3
Bristol-Myers	4444.0	531.4	261.7
Johnson & Johnson	6421.3	613.7	471.1
Eli Lilly	3270.6	517.6	369.8
Merck	3547.5	539.9	426.3
Pfizer	4024.5	579.7	286.7
Schering-Plough	1927.1	192.6	175.4
SmithKline Beckman	3256.6	514.4	309.6
Squibb	2041.7	226.6	165.7
Syntex	948.5	150.0	127.7
Upjohn	2017.2	203.2	284.1
Warner-Lambert	3200.1	−315.6	208.2

Note: Figures for sales, profits and R & D refer to all activities, not just those related to biotechnology.
Source: Corporate accounts

TABLE 4.8 REVENUES, PROFITS AND RESEARCH AND DEVELOPMENT EXPENDITURES FOR SELECTED CHEMICAL COMPANIES PARTICIPATING IN THE BIOTECHNOLOGY MARKET, 1985
(in millions of $)

Company	Revenues	Profits	R & D exp.
American Cyanamid	3536.1	120.2	250.6
Celanese	3046.0	178.0	103.0
Dow Chemical	11537.0	58.0	547.0
Du Pont	29483.0	1118.0	1144.0
W R Grace	5193.0	127.8	92.0
Monsanto	6747.0	−128.0	470.0
Rohm & Haas	2051.0	141.0	124.0
Union Carbide	9003.0	−599.0	275.0

Note: Figures for sales, profits and R & D refer to all activities, not just those related to biotechnology.
Source: Corporate accounts

TABLE 4.9 REVENUES, PROFITS AND RESEARCH AND DEVELOPMENT EXPENDITURES FOR SELECTED OTHER COMPANIES PARTICIPATING IN THE BIOTECHNOLOGY MARKET, 1985
(in millions of $)

Company	Revenues	Profits	R & D exp.
Campbell Soup	3988.7	197.8	34.5
Corning	1690.5	107.6	87.5
Dart & Kraft	9942.3	466.1	85.9
Eastman Kodak	10631.0	332.0	976.0
Perkin-Elmer	1304.6	82.1	99.0
Ralston-Purina	5863.9	256.4	48.5
Weyerhauser	5205.6	200.1	44.1

Note: Figures for sales, profits and R & D refer to all activities, not just those related to biotechnology.
Source: Corporate accounts

1982, its cloned insulin was approved for sale in the USA and the UK, thereby becoming the first recombinant-DNA drug to reach the market. Its human growth hormone was approved for sale the next year.

By 1979, based on its success in garnering research contracts with other companies, Genentech became the first biotech startup to show a profit. In 1980, the company made history again by becoming the first biotech company to go public, raising nearly $40 million. Over the years, Genentech's focus has remained human health care, mainly therapeutic drugs and hormones, although some diversification has occurred in the areas of diagnostics, instruments, agriculture and food processing. Like most other startups, the company has used a variety of methods to raise funds, including public placements, joint ventures and research contracts. Besides Eli Lilly, the company has contracts and research ventures with numerous foreign drug and chemical companies, including Hoffmann-La Roche, which has rights to produce the company's alpha- and beta-interferon products. Genentech was also an innovator in the use of R & D limited partnerships as a vehicle for raising research funds.

In 1985, Genentech earned nearly $90 million in revenues from product sales, royalties, contract revenues and investment income. However, like most biotech startups, very little of this income was derived from product sales (only about 6 per cent).

The company's management has announced a goal of $1 billion in sales by 1990. Such a goal will be impossible to achieve unless the company begins to utilize its technology to manufacture and market its own products rather than developing products under contract or licensing its technology. To this end, the company bought out two of the limited partnerships that fund its research. The company has also pursued a dual-track strategy of either retaining US rights to manufacture and market drugs and other technology it develops or developing products on its own. Among the most prominent products Genentech is pursuing without partners is a series of vaccines, such as those for the treatment of herpes, hepatitis and AIDS.

Cetus is the second largest publicly-held biotechnology company in the USA. Although it was formed five years earlier than Genentech, in 1971, its revenue base is much smaller. In 1985 total revenues for the firm were about $50 million. Cetus, like Genentech, went public in 1980, following Genentech's lead by just a few months. However, its stock offering was far larger, and $120 million was raised. Like Genentech, Cetus plans to become a fully integrated pharmaceutical company using biotechnology-based technology instead of more traditional drug manufacturing techniques. It too has concentrated on human health care products, although there is little overlap in research at the two companies. Also, unlike Genentech, Cetus' research exhibits somewhat more balance between therapeutic products and diagnostic products, with other research being conducted for instruments, food processing and agricultural applications.

Cetus' human health care research has mainly been conducted with funds raised through an R & D limited partnership, the Cetus Healthcare Limited Partnership (CHLP). Only the company's beta-interferon product has been developed with funds from non-CHLP sources among its therapeutic and diagnostic products. For other products, CHLP retains right of manufacture and distribution worldwide, although there are buyout provisions that Cetus may exercise. For non-human health care products, however, Cetus has traditionally sought funding from other partners, such as W R Grace in agriculture (e.g. for genetically improved plants and microbial crop treatments) and Nabisco in food processing (for development of food processing techniques and new food additives).

Besides Genentech and Cetus, there are several other important biotech startup companies:

Amgen. In 1985, Amgen was the fourth largest public biotechnology startup ranked by revenue (excluding Hybritech, which was acquired by Eli Lilly). It has focused on developing human health care products, and, like most other startups, Amgen's ultimate objective is to become a fully integrated health care products manufacturer. Among the products it has developed are IL-2, hepatitis B vaccine and gamma interferon (all to the clinical trials stage). Amgen has also concentrated on developing diagnostic tests and 'growth factors' for promoting tissue growth after wounds. The company is diversifying into animal health care and speciality chemicals (including a venture with Kodak to develop nutritional supplements).

Applied Biosystems. This company was incorporated in 1980 and specialises in providing instruments, reagents, biochemicals and other consumables to biotechnology laboratories. Included among its products are computer-controlled, chemical robots for analysis or synthesis of protein and DNA. This and other technologies developed by the company are particularly relevant to the development and production of semi-automated or fully automated diagnostic systems. Applied Biosystems is the third largest biotech startup ranked by revenue and is the largest such firm which derives the majority of its revenues from product sales (as opposed to research contracts and/or investment income). It supplies its products to many leading biotech companies.

Biogen. Biogen was founded in 1978 as a US-Swiss venture including scientists from several nations. Corporate investors included multinational firms, but most of the firm's funds were raised in the USA. Biogen has focused its research on developing human health care products, including anti-cancer and anti-inflammatory products, with additional research directed at developing products for the treatment of heart attacks, strokes and viral infections and development of growth stimulants and vaccines. (See also Chapter Six.)

Bio-Response. This firm was founded in 1972 and is primarily engaged in the production of cell cultures under contract. It has developed two significant technologies – a mass culturing technique and a series of cell isolation techniques. The company's revenue base is quite small, and much of its past contracts were derived from a single customer – Ciba-Geigy Ltd. However, the company is trying to diversify and hopes to use its technology to develop and produce diagnostic test kits and therapeutic drugs.

Centocor. Centocor was founded in 1979. Currently it manufactures a number of diagnostic products for commercial sale including in vitro cancer diagnostic tests using monoclonal antibodies. They have been approved for sale in Europe and Japan (foreign countries accounted for about 75 per cent of sales volume in 1985), and application for approval has been made in the USA. The company also manufactures an FDA-approved hepatitis B test and is evaluating a test for diagnosing AIDS (both tests are licensed to Du Pont). Centocor is also using its monoclonal antibodies expertise to develop diagnostic imaging tests for cardiovascular disease and human therapeutic products (e.g. gastrointestinal cancer and gram-negative bacterial infections).

Chiron. Chiron was formed in 1981 and went public in 1983. The company is developing a wide range of biotech products, including vaccines, hormones and growth factors, therapeutic enzymes and diagnostic products. At this time the company commercialises its products primarily through licensing or supplying semi-finished products to other manufacturers. Despite its small size (revenues were only about $7 million, excluding investment income in 1985), Chiron recently gained recognition through FDA approval of its hepatitis B vaccine. This vaccine was a major focus of research at several biotech startups, and FDA approval for Chiron has allowed it to leap ahead of the pack. Merck & Co. will produce the vaccine. Chiron is also hard at work on an AIDS vaccine, but its future plans envisage continuation of product licensing for this and other low volume, niche products.

Damon Biotech. Damon Biotech's primary business is contract production of monoclonal antibodies for use by pharmaceutical manufacturers, diagnostic test producers and research laboratories. The company also has patent rights to technologies for producing monoclonal antibodies and genetically engineered proteins. Eventually, the company plans to use its proprietary technologies to manufacture and market various human health care products.

Diagnostic Products. This company specialises in manufacturing diagnostic kits, which are distributed in more than 70 countries. The kits use monoclonal antibody technology and have been developed for diagnosis of thyroid disorders, anaemia, fertility for pregnancy, allergies, cancer and management of several hormone-related disorders/imbalances. The firm is working on simplified versions of many of its tests for use in physicians' offices and in the home. With revenues of more than $22 million in 1985,

Diagnostic Products is one of the largest biotech startups. In addition, virtually all of the company's revenues are derived from product sales.

DNA Plant Technology. This firm, founded in 1981, specialises in agricultural applications of biotechnology using cell- and tissue-culture technology rather than recombinant-DNA to develop improved plants. The company does not conduct contract research, but prefers to enter into agreements with partners that provide for joint venture, licensing or royalty arrangements for products/technologies developed. Among its partners are Kraft (vegetable snacks), United Fruit (cloning for improved oil palm trees) and Campbell Soup (improved tomato varieties).

Genex. Genex was founded in 1977. The company originally intended to use its expertise in the development of bioprocesses to become a producer of chemicals. As part of this strategy, the company developed a proprietary process for manufacturing L-phenylalanine, a key component in the production of the sweetener aspartame. In 1983, Genex signed an agreement with Searle (the maker of aspartame) to manufacture L-phenylalanine, and, in late 1983, Genex brought a new production facility on line in Kentucky. This facility was expected to provide the major share of the company's revenues. However, in 1985, Searle terminated the contract, forcing Genex to close the facility and resulting in major losses. For this reason, the company has rethought its strategy and now plans to concentrate on providing biotechnology services, not the development and marketing of proprietary chemicals.

Molecular Genetics. This firm has focused on biotechnology products for the agricultural sector. Its first product was a monoclonal antibody treatment for calf scours, introduced in 1983. The company is also developing a variety of animal vaccines, often under research contracts sponsored by major corporate partners (such as Eastman Kodak). Molecular Genetics is also conducting research aimed at developing improved plant strains (e.g. herbicide-resistant corn with American Cyanamid) and improved animal feed additives.

Monoclonal Antibodies. This company was founded in 1979. It develops, produces and markets diagnostic products which utilise monoclonal antibody technology. The company's best-known product is its line of pregnancy tests, originally developed for the professional market, but later reconfigured for in-home use. These are marketed by Ortho Pharmaceutical Company (a subsidiary of Johnson & Johnson) under the 'Advance' brand-name. Monoclonal Antibodies sells components for this test to Ortho, which provides other components and packages and markets the finished product to consumers. Monoclonal Antibodies has also developed human ovulation tests and horse pregnancy tests for sale. Products under development include a variety of human and animal fertility tests and a number of human infectious disease tests.

4.5 Funding of R & D

There have been two major sources of funds for biotechnology research and product development in the USA: public and private funding. Public funding has been dominated by support from various federal agencies and has largely been confined to grants to educational institutions and public and quasi-public labora-tories (e.g. the so-called 'national' laboratories that are run by private-sector companies under contracts with the federal govern-ment). Public funds have largely been used in support of basic research into biotechnological processes/techniques and into defence-related issues.

Private funding, by contrast, has been largely used to raise capital for startup biotechnology companies and for new research at large companies (such as Monsanto and Eastman Kodak). While some of these funds are earmarked for basic research, most are designed to develop processes and products for commercial applications.

PUBLIC SOURCES

No comprehensive census of federal funding in support of biotechnology research has ever been undertaken. However, the US General Accounting Office and other sources have attempted to assemble some information on federal funding. By far the largest source of funding is the National Institute of Health, which accounts for more than 80 per cent of federal biotech research support. NIH directs its funds to two major areas of research – basic research into biotechnology processes and techniques, and basic research aimed at gaining a better understanding of fundamental biological processes. The latter research area includes the study of genetics, immunology and molecular and cell biology.

Among the other major federal contributors to biotech research are the Department of Defense (which splits its research budget relatively evenly between basic and applied research), the National Science Foundation (almost entirely basic research), and the Department of Energy (almost 80 per cent basic research). The Defense Department research programme is aimed at developing possible biological warfare weapons, while the Energy Depart-ment's research programme focuses on developing biological processes relevant to energy production (e.g. processes for converting wood to liquid or gaseous fuels).

TABLE 4.10 FEDERAL FUNDING OF BIOTECHNOLOGY RESEARCH, BY AGENCY, 1986 (in millions of $)

Agency	Amount for fiscal year 1986
Department of Agriculture	74.1
Department of Defense[1]	106.0
Department of Energy	55.9
Environmental Protection Agency	3.4
Food and Drug Administration[1]	3.5
National Aeronautics and Space Admin.	6.4
National Bureau of Standards	3.3
National Institutes of Health	1910.0
National Oceanic and Atmospheric Admin.[1]	2.2
National Science Foundation[1]	85.0
US Agency for International Development	14.3
Total	2264.1

Note: [1] Estimated
Source: US federal agencies

Another federal agency with a relatively large biotechnology R & D budget is the Department of Agriculture. Table 4.11 breaks down research funding, by type of project, for various segments of the Department of Agriculture. Biotech research funded by USDA is carried out by in-house staff and through contracts with non-profit institutions (e.g. universities) and with private companies (such as Genex and Genentech).

**TABLE 4.11   FUNDING OF BIOTECHNOLOGY RESEARCH BY
THE US DEPARTMENT OF AGRICULTURE BY
PROGRAMME AREA, 1986**
(in millions of $)

Agency	Amount for fiscal year 1986
Agricultural Research Service (ARS)	
Maintaining quality of natural resources	0.23
Improving crop production and protection	
Germplasm diversity	0.29
Modify germplasm	2.64
Crop productivity/quality	2.89
Plant protection	4.38
Improving animal production and protection	
Genetics and reproduction	1.69
Diseases	6.17
Other (nutrition, insect control)	0.91
Improving product quality	7.36
Improving human nutrition	0.05
Total ARS	26.60
Forestry Service	
Tree genetics/improvements	1.20
Resource utilisation	0.20
Total Forestry Service	1.40
Cooperative State Research Service (CSRS)	
Competitive research grants	
Animal growth and development	3.40
Animal molecular biology	4.40
Plant growth and development	2.20
Plant molecular biology	4.10
Response to abiotic stress	2.00
Response to biotic stress	2.10
Biotechnology research included in other grants	9.50
Biotechnology funding from other CSRS programmes	16.60
Total CSRS	46.10
Total Department of Agriculture	74.10

Source: US federal agencies

PRIVATE SOURCES

By 1986, approximately $3.5 billion in research money had been
raised by publicly-held US biotechnology companies. More than
70 per cent of this figure represented equity funds, with contract

revenues (mainly from large chemical, pharmaceutical and agricultural processing companies) the second largest source.

TABLE 4.12   SOURCES OF PRIVATE FUNDS FOR US PUBLIC BIOTECHNOLOGY FIRMS, 1985

Source	$ million
Equity	2500
– Public placements	1900
– Private placements	600
Research revenues	900
– Contracts	700
– R & D limited partnerships	200
Debt	100
Total funding	3500

Source: Montgomery Securities

This $3.5 billion figure includes funds raised by only 65 publicly-held biotechnology companies. It does not include either internal funding of biotech research by large, diversified firms or funding of the more than 250 privately-held US biotech firms. Although some of the research money spent by large, diversified firms is captured by the contract revenues figure in Table 4.12, these firms have invested heavily in internal research as well.

Tables 4.13 and 4.14 present information on the type of research conducted with the $3.5 billion raised by biotech startups. As the tables show, much of the research effort is directed at developing therapeutic treatments for cancers and cancer-related diseases, with therapeutic drugs for other diseases and diagnostic products also accounting for a large share of research funds. Among the technologies used in the biotech industry, recombinant DNA techniques have received the largest share of funds.

4.6   Outlook

MARKET PROSPECTS

Federal regulatory policy toward biotechnology has been highly controversial, particularly as products have moved beyond the

TABLE 4.13 RESEARCH FUNDS RAISED BY PUBLICLY-HELD BIOTECH STARTUPS, BY APPLICATION CATEGORY TO 1985

Application	Percent of all funds raised
Cancer therapeutics	43
Other therapeutics	19
Diagnostics	13
Crop improvement	12
Speciality chemicals	4
Agricultural chemical	4
Research laboratory supplies	3
Animal health	2

Source: Industry estimates

TABLE 4.14 RESEARCH FUNDS RAISED BY PUBLICLY-HELD BIOTECH STARTUPS, BY TECHNOLOGY USED TO 1985

Application	Percent of All Funds Raised
Recombinant DNA	62
Hybridoma/monoclonal antibody	30
Other[1]	8

[1] Includes fermentation, tissue culture, etc.
Source: Industry estimates

laboratory to the testing and commercialisation stages. Three federal agencies have jurisdiction over various aspects of the biotech industry – the Environmental Protection Agency (EPA), the Food and Drug Administration (FDA) and the Department of Agriculture (USDA). Although policy statements published by the federal government in June of 1986 in the Federal Register have helped somewhat to clarify the role of each agency, there remain some disputes over the lines of demarcation between the jurisdictions of each agency. (See also Chapter Seven.)

Of particular concern to the industry has been the inadequate overview exercised in the past by the federal government and the inconsistent application of standards for granting approval of tests and/or commercial utilisation for biotech products. As a result, recent years have been marked by both major advances and major setbacks for the biotechnology industry.

On the plus side, the industry has gained approval for commercial use of some important biotech products. For example, in late spring 1986, the FDA approved commercial use of human alpha interferon produced by Schering-Plough (using technology developed by Biogen) and Hoffmann-La Roche (using technology developed by Genentech) for the treatment of hairy cell leukaemia. It was the first time a biotech-developed therapeutic drug had been approved for commercial use in humans. Also, in June 1986, Ortho Pharmaceutical received FDA approval for use of mono-clonal antibodies to reverse the rejection of newly transplanted kidneys, and Rohm & Haas received approval to field test genetically altered tobacco plants (which are resistant to leaf-eating caterpillars).

On the minus side of the ledger, there remains considerable controversy over planned field tests of genetically altered bacteria that would help plants resist frost. The concern over these tests was heightened when the bacteria's developer, Advanced Genetic Sciences, revealed that some bacteria may have been prematurely released into the environment as a result of sloppy test procedures. Other problems with federal regulatory policy were apparent when the USDA first approved the sale of a genetically altered vaccine for swine and then suspended its approval in April 1986. Later that month the vaccine was reapproved but challenged by activists who demand tighter federal regulation of biotech products.

Another area of some concern to participants in the biotech industry is the degree of patent protection that will be granted to their products and processes. Although patent protection was considerably strengthened by a 1980 Supreme Court ruling that genetically-altered life forms were patentable, several patent suits still outstanding could have significant impact on the scope of patent protection provided to the industry. (See Chapter Eight.)

INDUSTRY PROSPECTS

Two major trends will mark the future of the biotechnology industry in the USA. First, rapid growth in revenues will continue as existing products are approved for the commercial market and new products are developed. Second, the industry will consolidate into fewer major participants, with many existing entrepreneurial, startup companies taken over (in whole or in part) by larger, diversified manufacturing firms. This consolidation process has already begun with several takeovers (or minority equity investments) of biotech startups by larger manufacturing firms. To some extent, these investments by traditional manufacturers are entirely defensive in nature, since biotechnological production processes show promise in displacing more traditional production techniques for products already marketed by these larger firms. In other cases, these investments represent attempts by manufacturers at diversification into a fast growing industrial sector.

As the industry grows over the next 10 to 15 years, the product mix (in terms of both type of products and the technology used to produce them) will necessarily change as well. At the present time, the industry's commercial sales are currently dominated by diagnostic products, developed using monoclonal antibody techniques. Also, most sales are for diagnosis of human diseases and conditions. Finally, a substantial proportion of the industry's current product revenue base (i.e. excluding equipment and R & D contracts) is derived from contract research.

Over the short run, the dominance of monoclonal antibody technology will be reduced, as more recombinant DNA therapeutics are approved for human, animal and environmental use. Also, as more products are brought to market, the importance of contract funding to the industry's revenue base will be reduced. There are five major factors (several of which are closely interrelated) which will facilitate the reduced proportion accounted for by contract revenues:

– As more products are brought to market, biotech firms will devote proportionally less resources to basic research.
– As biotech firms gain more experience as manufacturers, they will be less willing to take on outside partners for development of technology. Instead, they will prefer to capture the benefits of product development internally.
– As biotech firms improve their cash flow through product sales, they will be able to generate a larger proportion of research funding needs internally.
– As larger, diversified manufacturing companies in the pharmaceuticals, chemicals, agribusiness and other sectors gain more experience in the field, they too will prefer to conduct more research using internal resources (instead of contracting out research to startups).
– Larger diversified manufacturers will, over time, take over

more of their current research contract 'partners', thereby adding to their internal biotechnology resources. The bases for these takeovers will be the desire to capture a larger share of the income stream from biotech product development/production and to gain access to proprietary technology/knowledge. One result will be to decrease the necessity of contracting out research, since adequate resources will be available in-house.

In terms of product revenue sales, compound growth will be in the order of 25–30 per cent per year for the 1985 through 2000 time period. This implies that product revenues (i.e. excluding contract revenues and equipment sales) will increase from about $400 million in 1986 to about $1.5–$2.0 billion by 1991 and to between $10 and $20 billion by the year 2000. More rapid growth will be recorded in the early years of the period, although growth will remain strong throughout the period. Growth forecasts are somewhat dependent on the speed of regulatory approval for biotech products in the USA (the world's largest market). Traditionally, government approval for similar products (such as human drugs) has occurred somewhat more slowly in the USA than in other nations.

The mix of biotechnology products on the US market will shift somewhat over time. Currently, diagnostic products for humans and (to a lesser extent) animals dominate the market. However, within 5 years drugs for humans and animals will account for over one-half of the market. Also, non-animal agricultural applications (such as improved seeds/plants and biotech pesticides and herbicides) should grow from virtually nothing in 1986 to about 20 per cent of the market by 1991.

By 1995, only about one-half of all biotech product sales will be derived from human and animal health care and diagnostics. Other important segments will be chemicals, seeds (for genetically altered plants), food processing/food additive products and pesticides/herbicides. Commercial application of biotechnology to fuels production, mining and electronics is unlikely to occur on a wide scale before 1995–2000.

5 Chapter Five

The Biotechnology Industry in Japan

5.1 Introduction

The background information and statistics for this review cover the period up to year 2000. Beyond that year developments are less easily predictable and often the subject of various intelligent guesses of varying quality. Developments in Japanese biotechnology up to 1990 can be reasonably well anticipated as progress reports of various ongoing research projects are made public and announcements of new new biotechnology products are made. Biotechnology has given rise to a new industry in Japan, barely ten years old. A great number of companies from various industrial fields have entered what is now called the 'bioindustry'.

As of 1985, the Japanese bioindustry market was estimated to be worth about $286 million, and is likely to increase to at least $600 million around 1990 and $86 billion by the year 2000. Few commercial biotech products have appeared so far. Current R & D efforts are focused on pharmaceuticals and foods. By the year 2000, the market for bio-pharmaceuticals is forecasted to be $18 billion, and the bio-food market $24.3 billion. Most other biotech markets will be well below $10 billion annually then.

More than 300 companies from various industrial fields make up Japan's bioindustry. About half of these form the core of the industry and a top-echelon group of 10–20 companies dominates. Total biotech R & D expenditure in Japan in 1985 was estimated at $820.8 million. The government share is about 34 per cent, and the Ministry of Education is the biggest government spender. The total number of biotechnology researchers is about 4000. Recombinant DNA and cell fusion technologies are the focus of research in the pharmaceutical industry, and bioreactor systems in the chemical and food industries.

5.2   Structure of the Bioindustry in Japan

MAJOR COMPANIES

Corporate participation in the bioindustry reads like a Who's Who of Japanese industry. While the total number of companies is not exactly known, it is estimated that more than 300 manufacturers from a wide variety of industries are involved. The 148 member companies of the Bioindustry Development Center (BIDEC) form the core of the industry. These numbers increase further if foreign-capital companies in Japan, and small- and medium-sized fermentation-related Japanese companies diversifying into biotechnology are included. According to the Ministry of International Trade and Industry (MITI), there were 2600 breweries, 1887 bean paste, and 2927 soy sauce manufacturers in 1983, totalling more than 7400 fermentation enterprises.

Classification of participants by industry is becoming increasingly blurred as more companies are branching out into related fields: food companies into pharmaceuticals and vice versa; textile companies into chemicals and the reverse. Biotechnology is becoming an industrial melting pot. The scenario for the next few years calls for pharmaceuticals, particularly physiologically active peptides, to be the nucleus of development, with leading companies in the field taking the lead in commercialisation, teaming up ad hoc with universities and research institutes and sometimes licensing technology from overseas firms, particularly small specialist US biotechnology companies or developing collaborative efforts on a national scale (see Table 5.6).

A closer look at the pharmaceutical industry shows a top layer of 12 big-name companies selling prescription drugs, trailed by a 12-strong group of second-tier companies, and followed by ten small-and medium-sized companies, making up a group of specialised pharmaceutical companies. Below these comes a group of five companies for which drugs are only one part of their business, but include at least two that rank top in the industry. Some seven foreign-owned and ten companies from the food and chemical industries are also in the race. A group of non-prescription drug manufacturers of various sizes (more than eight at least) are at the lower end. About half of the above 60-odd companies are in the process of developing new original drugs with varying degrees of success. Estimates in the industry set the number of companies capable of successful applications in the near future at between ten and 20.

TABLE 5.1   CORPORATE CATEGORIES IN THE PHARMACEUTICAL INDUSTRY

Pharmaceutical corporations	Japanese-owned companies specialised in pharmaceuticals	12 Major corporations selling prescription drugs	Takeda Chemical Industries, Sankyo, Fujisawa Pharmaceutical, Shionogi, Tanabe Seiyaku, Eizai, Yamanouchi Pharmaceutical, Chugai Pharmaceutical, Banyu Pharmaceutical, Dainippon Pharmaceutical, Yoshitomi Pharmaceutical Industries.
		Second-tier corporations selling prescription drugs	Nippon Shinyaku, Otsuka Pharmaceutical, Taiho Pharmaceutical, Mochida Pharmaceutical, Toyama Chemical, Torii, Kyorin Yakunin, Ono Pharmaceutical, Nikken Chemicals, Tokyo Tanabe, The Green Cross Corp., Kaken Chemical.
		Small and medium-sized corporations selling prescription drugs	Fuso Pharmaceutical Industries, Sanwa Kagaku Kenkyusho, Morishita Pharmaceutical, Nippon Chemiphar, Kissei Pharmaceutical, Teikoku Hirmone Mfg, Hokuriku Seiyaku, Nihoniyakuhinkogyo, Funai Pharmaceutical, others.
		5 companies with a pharmaceuticals division	Meiji Seika Kaisha, Kyowa Hakko Kogyo, Sumitomo Chemical, Nippon Kayaku, Toyo Jozo.
	Others	Foreign-owned companies	Taito Pfizer, Nippon Hoechst, Nippon Merck Banyu, Japan Upjohn, Ciba-Geigy (Japan), Sandoz Pharmaceutical, Beecham Yakuhin K. K., others.
		Non-pharmaceutical companies	Mitsubishi Petrochemical, Mitsubishi Yuka Pharmaceutical, Mitsui Toatsu Chemicals, (Mitsui Pharmaceutical), Teijin, Toray Industries, Mitsubishi Chemical Industries, Suntory, Showa Denko K. K., Asahi Chemical Industry, Ajinomoto, Kanebo (Kanebo Pharmaceutical), others.
Large, medium-sized and small corporations selling non-prescription drugs			Taisho Pharmaceutical, Kowa Shinyaku, Tsumura Juntendo, SS Pharmaceutical, Rohto Pharmaceutical, Santen Pharmaceutical, Wakodo, Hisamitsu Pharmaceutical, others.

Source: Industrial Bank of Japan

79

TABLE 5.2   MAJOR INDUSTRIAL COMPANIES AND R & D (BIOTECH) BUDGETS, 1984

Company	Main sector(s)	Total sales (yen) (billion)	R & D in biotechnology Budget (yen) (million)	R & D in biotechnology Researchers (no.)
Kirin Brewery	Brewing	1042	300	20
Suntory	Beverages, food	803	n.a.	n.a.
Mitsubishi Chemical	Chemicals	747	4000	200
Asahi Chemical	Chemicals, textiles, construction, food	707	3000	200
Sumitomo	Chemicals	659	n.a.	80
Toray	Fibres	613	n.a.	n.a.
Takeda	Pharmaceuticals	478	911	46
Snow Brand	Dairy	429	1600	78
Ajinomoto	Food	423	n.a.	n.a.
Teijin	Fibres, plastics	413	600	50
Showa Denko	Chemicals	411	1000	50
Mitsui Toatsu	Chemicals	401	2000	130
Meiji Milk	Dairy	352	1500	50
Toyobo	Textiles	349	1500	50
Unitika	Textiles	247	1000	50
Kyowa Hakko Kogyo	Chemicals, pharma., food, beverages	224	9200	430
Toyo Soda	Chemicals	224	1500	50
Shionogi	Pharmaceuticals	177	499	45
Tanabe	Pharmaceuticals	143	1140	40
Kikkoman	Food, beverages	136	450	21

Note: n.a.= breakdown of R & D for biotechnology not available.
Source: Company reports

Biotechnology R & D expenditures as a percentage of annual sales may serve as a good indication, though not the best, of who will be calling the shots in the bioindustry. Among some 20 leading companies from various industries engaged in biotechnology research, the average amount annually spent is about 0.5 per cent of sales. (It varies up and down from industry to industry.) The figure may appear low considering the advances made by Japanese companies in recent years. One plausible explanation might be that biotechnology, a hands-on technology, is more dependent on brains and top-quality researchers than on the actual amount of money spent. Another factor of uncertainty is that some of the leading companies decline to reveal their actual R & D budgets for fear of giving away too much to competitors. Another complexity is that 'bought in' research in the form of product licensing deals may appear under different headings.

One industry leader, Kyowa Hakko Kogyo, a top-notch company in fermentation technology, is spending more than 4 per cent of its $1.3 billion annual sales on biotech R & D. This is by far the largest share among the top 20. Although pharmaceutical companies will be most prominent in the early years, larger Japanese companies from the food, agricultural and industrial sectors will emerge later. Another big spender is Mitsubishi Chemical Industries, which budgets about 1.3 per cent or close to $60 million of its annual $4.3 billion sales on biotech R & D. This is by far the largest single biotech R & D budget in the industry. Kirin Brewery, though with over 60 per cent of the domestic Japanese beer market and annual sales of about $6 billion, reports only $1.7 million as biotech expenditures, a mere 0.03 per cent, according to BIDEC. Asahi Chemical, another giant in the chemical industry with $4 billion in annual sales, spends some $17 million annually on biotech R & D. On average, many companies in pharmaceuticals, chemicals and foods, save the very few at the absolute top, seem to work with $10 million annual R & D budgets and between 15 and 30 R & D staff. It must be remembered, however, that in many cases the cost of R & D is being shared with foreign companies when it comes to major projects.

MAJOR MARKET SECTORS

The bioindustry in Japan is still in its infancy, approaching a transition point between inception and preparation for growth. The actual size of the market is difficult to grasp and assess with greater accuracy for several reasons intrinsic to the nature of biotechnology, a general purpose basic technology with a wide range of applications in several industrial fields.

The impact of biotechnology will not be fully felt until well into the next century. Up to the year 2000, its impact can be characterised by sporadic new product developments for a limited range of application fields, notably pharmaceuticals. Estimates in the industry set the current size of the market at about $285.7 million (¥50 billion), but this figure is greatly dependent on the

TABLE 5.3   JAPAN'S INTERNATIONAL COOPERATION IN BIOTECHNOLOGY

Bilateral Projects with Advanced Nations

Country	Agreement	Project	Implemen-tation	Theme	Method
United States	Japan-US Energy Research and Development Agreement (1980–)	Biomass conversion	1981		● Information exchange ● Exchange of specialists
West Germany	Japan-W. Germany Science and Technology Agreement (1974–)	Biology and medicine panel	1975	(1) Micro-organisms for waste treatment (2) Fermentation technology (3) Artificial organs	● Information exchange ● Conference and exchange of specialists ● Exchange of views by specialists
France	Japan-France Science and Technology Agreement (1974–)	Biotechnology	1982	(1) Engineering application technologies for enzymes (2) Technology for energy from biomass conversion	● Information exchange
Sweden	AIST-STU Cooperation Council	Biotechnology	1982	Fermentology	● Information exchange ● Exchange of views by specialists and joint research
		Lignin	1983	Chemistry of lignin	

International Collaboration Projects in Biotechnology Established as a Result of Discussions at Summit Conferences

Project	Leader	General description
Food technology	France, UK	Food-processing technology, safety evaluation standards, and the role of industrial countries in assisting developing countries.
Marine culture	Canada	Establishment of R & D planning group to reinforce production of marine products in cold-water environments, etc.
Biotechnology	France, UK	Information exchange for developments in biotechnology, assistance for researchers, etc.

Source: MITI (both tables)

82

products included. There are both lower and higher estimates. Another element of uncertainty is the number of companies included in sample surveys conducted by government agencies, media, and trade and industry associations. Common to all these surveys is the vast market potential forecast for year 2000.

A 1982 survey covering 39 separate industrial fields by the Ministry of International Trade and Industry (MITI) forecast the bioindustry market scale in year 2000 at between $24 billion and $39 billion with pharmaceuticals alone taking a share of 35–50 per cent. Another more recent survey conducted by a leading financial daily (the Biotechnology R & D Movements Survey by Nihon Keizai Shimbun and Nikkei-McGraw-Hill), covering 268 corporations, predicts a total market value of $47.2 billion by 2000.

Perhaps the most authoritative market forecast has been made by the Bioindustry Development Center (BIDEC) at the Japanese Association of Industrial Fermentation (see Table 5.4).

TABLE 5.4   JAPAN'S BIOINDUSTRY – MARKET SCALE FORECAST YEAR 2000

Industry sectors	Total output $ billion	Biotechnology $ billion	Share %
Agriculture	66.7	8.0	12
Livestock; dairy	11.3	2.7	24
Forestry	4.9	0.05	1
Fisheries	22.5	0.7	3
Foods	105.5	24.3	23
Paper & pulp	10.3	0.5	5
Basic chemicals	66.9	8.7	13
Fine chemicals	38.4	6.1	16
Pharmaceuticals	45.0	18.0	40
Agri-chemicals	2.7	0.8	30
Natural resources	31.9	1.9	6
Energy	132.2	2.6	2
Utilities	34.0	7.8	23
Electrical machinery	114.9	3.4	3
Sub-total	687.2	85.6	12.5
Biotech R & D support Industry	39.2	3.4	8.7
Total	726.4	89.0	12.3

Note: Exchange rate: $1=¥175
Source: Bioindustry Development Center (BIDEC)

The total output of 14 major bioindustry sectors is forecast to be worth $687.2 billion by 2000. The share of biotechnology (i.e. the output resulting from biotechnology production processes) is estimated at $85.6 billion. A more detailed breakdown into 51 separate industrial fields, including the not yet clearly defined biotech R & D support industry, increases the total output value to $726.4 billion, and the share of biotechnology to $89 billion. Thus, total bioindustry output will represent a 12.4 per cent share of Japan's Gross Domestic Product, estimated at $5879 billion (¥ 1,028,000 billion) by the year 2000.

Pharmaceuticals are currently the most promising field for biotechnology. The focus is on physiologically active substances such as interferon, interleukin-2 (IL-2), tumour necrosis factor (TNF), tissue plasminogen activator (TPA), hepatitis B vaccine and human growth hormones. Each has a potential billion dollar market. Commercial products such as insulin and beta-interferon have already appeared on the market, and alpha-interferon approval is imminent. The annual market for each is estimated at about $28.6 million at present. An additional ten products will possibly be marketed before or by 1990. TPA alone may generate $570 million a year then. The same may hold for gamma-interferon. Monoclonal antibodies as anticancer drugs and clinical diagnosis form another big market, currently valued at about $3 million yearly, and expected to grow into an annual $40 million market by 1990, according to BIDEC.

The domestic Japanese market for pharmaceuticals is currently worth about $23 billion annually, and is forecast to grow to about $45 billion by 2000. Biotechnology drugs are then forecast to account for 40 per cent or $18 billion. This is a higher share than in any other industry.

The food industry is already well versed in fermentation techniques and is now applying biotechnology to produce foods and drinks. Few marketable products have appeared so far, but dairy products and food raw materials are the most promising fields. Production costs are the major concern. The current focus is on increasing productivity in amino acid production through the use of bioreactors and immobilised enzymes. The annual amino acid market is worth about $1.5 billion at present, and is likely to grow substantially once production technology is fully developed.

New sweeteners form another big future market, and include sugars and non-sugars. The total market value is about $114.3

million now of which some 10 per cent falls to biotechnology. Isomerised sugar is the leading product and may have a $171.4 million market by 1988. The total sugar market value is estimated at $6.2 billion by the year 2000, and biotechnology may have a 10 per cent share by then. According to BIDEC, the total Japanese food industry output will be valued at $105.5 billion by the year 2000, and biotechnology will account for about 23 per cent or $24.3 billion.

The chemical industry is in a similar situation with no commercially viable products in sight for the near future. Reportedly, only one fine chemical product has a commercial potential with bioreactor technology.

In the fine chemicals fields, surfactants and cosmetics have seen some advances in biotechnology. Some bio-cosmetic products are already on the market such as lipsticks and some liquids. Despite the fact that a 5–10 year lead time is necessary for full commercialisation, bio-cosmetics already have a $240 million market. BIDEC envisages a ten-fold increase to $2.6 billion by the year 2000, including surfactants. It is also expected that biological raw materials for cosmetics, natural colour and aroma will have bright commercial prospects, forming the major part of the biotechnology share in this field.

By the year 2000, the total output of basic chemicals is expected to amount to $66.9 billion with biotechnology accounting for $8.7 billion or 13 per cent. Fine chemicals will have a $38.4 billion market, and biotechnology a share of 16 per cent or $6.1 billion.

Biotechnology has a huge long-term potential in agriculture, but real advances have proved slow in coming. Plant breeding technology has the leading edge at present. The market for various seeds is currently at the centre of interest. This market is valued at $1.2 billion, 25 per cent (including rice, grains, soy beans and potatoes) controlled by central and regional government authorities, and 75 per cent (including vegetables, flowers, feeds etc.) distributed by private companies. By 1990, the market is expected to be worth about $1.7 billion. The regulated, government-controlled segment is predicted to decline to about 21 per cent in that year as a result of stepped-up efforts by private companies. One leading company is expecting annual sales of about $2.9 million from biotechnology-bred flowers.

The value of other agricultural markets is still uncertain, but BIDEC forecasts total agricultural output at $66.7 billion by the year 2000, and a biotechnology share of 12 per cent or $8.0

billion. Livestock and dairy farming will grow to an $11.3 billion market with biotechnology accounting for about 24 per cent or $2.7 billion.

Parallel with the growth of the bioindustry, a peripheral industry, loosely defined as the biotechnology R & D support industry, comprising manufacturers of various biotechnology equipment such as DNA sequencers and synthesisers, cell sorters, peptide sequencers and various reagents has begun to mushroom. The exact size of this market is not fully known, but industry sources estimate it to account for about 10 per cent of the total bioindustry market at present, somewhere in the neighbourhood of $30 million annually. This share is likely to grow as the bioindustry grows.

5.3   R & D Funding and Organisations

Biotechnology research is rapidly coming of age in Japan with government-run R & D institutions, universities and leading private corporations from several industries as the major players. Private corporations have the edge in applications research for commercialisation, backed by basic research in universities and actively supported by government institutions through general R & D guidelines, project initiation and promotion, and to a lesser extent, funding.

PUBLIC SUPPORT

The main government bodies involved are the Ministry of International Trade and Industry (MITI), and its Agency for Industrial Science and Technology (AIST); the Ministry of Health and Welfare (MHW); the Ministry of Agriculture, Forestry and Fisheries (MAFF); and the Ministry of Education (MOE). Government involvement, which provided a valuable impetus in the early R & D stages, is now largely concerned with the establishment of safety guidelines, compilation of statistics, definition and standardisation of terminology, basic research funding, gene resource collection and systematic storage, database construction, industrial location policy, and international research co-operation.

The basic framework for biotech R & D efforts in Japan is set by MITI's and AIST's 'Research and Development Project of Basic

System for "Research and Development Project of Basic Technologies for Future Industries"

Center for the Project Promotion
(Ministry of International Trade and Industry)
General Manager

Staff Members
└ Committee for Basic Policy
  Committee for Technological Policy

Secretariat

Committee for Research Coordination
- Coordinator

Industrial Technology Council
- Committee for Research and Development Project of Basic Technologies for Future Industries
- Subcommittee for Biotechnology

Agency of Industrial Science and Technology (Project headquarter)

Committee for Research Evaluation

Cooperation with Academia

(Members in charge of Research)

National Research Institute

Research Contractor
(Private Sectors)

- Fermentation Research Institute
- Research Institute for Polymer and Textiles
- National Chemical Laboratory of Industry

- Research Association for Biotechnology
(14 Private Companies)

note:
☐ Member of Biotechnology Research

Technologies for Future Industries', initiated in 1981, which has the following objectives for the biotechnology part:
— Process simplification and lower energy consumption in the chemical industry.
— Development of technologies to solve the issues of exhaustion of natural resources and shortage of food supply.
— Large-scale production of pharmaceutical and food products.

With these objectives, the projects aim at developing basic technologies essential for the establishment of 'next-generation' industries, through a tripartite collaboration among government, industry and academia, and to develop and commercialise unique domestic biotechnologies, and applications thereof, to reduce reliance on imported technologies and to strengthen Japan's future economic security.

Under this umbrella project, the Biotechnology Forum was established in 1980 to promote biotechnology and to study the basic policy for implementing joint research efforts. As a result, in September 1981, the Research Association for Biotechnology was organised by 14 member companies for research in the following areas:
Bioreactor technology using micro-organisms or enzymes to develop new chemical processes for an energy-saving and pollution-free chemical industry;
Large-scale cell cultivation technology to replace serum in the media for various cell lines, making submerged cultures possible to reduce cost and time of purification;
Recombinant DNA technology to obtain hormones and enzymes for use in bulk production of chemicals.

In April 1983, the Bioindustry Development Center (BIDEC) was established by the Japanese Association of Industrial Fermentation (JAIF) to promote and coordinate industrial biotechnology R & D. As of March 1985, BIDEC had a 148-strong corporate membership. Supported by MITI, universities and its corporate members, BIDEC has become the focal point of biotech R & D activities and information in Japan, engaged in organising overseas study missions and publishing a newsletter to promote international cooperation in biotech R & D. A growing number of biotech R & D associations has been formed recently, partly through funds provided by the Japan Key Technology Center, jointly run by MITI and the Ministry of Posts and Telecommunications (see Table 5.6). Also recently, MAFF and the Ministry of Finance have set up a joint fund to promote R & D projects in the private sector.

TABLE 5.6 MAJOR R & D COOPERATIVE GROUPS IN BIOTECHNOLOGY

Groups	Participants	Activities
Protein Engineering Research Institute	Toray, Mitsubishi Chemical, Kyowa Hakko, Takeda Chemical, Toa Nenryo Kogyo	Research on protein structure and application of protein engineering techniques in industrial production.
M & D Research Co	Daicel Chemical, Meiji Seika	Synthesis and genetic engineering of peptides and research on assay methods of peptide activity.
Japanese Research and Development Association for Bioreactor Systems in Food Industry	54 firms, including Meiji Seika, Ajinomoto	Research on bioreactors, biosensors and enzymes for food production.
Research Association for Biotechnology of Agricultural Chemicals	16 firms, including Hokko Chemical and Kyowa Hakko Kogyo	Basic research on agricultural chemicals and development of new products using cell culture technique.
Human Science Research Foundation	About 125 firms in chemicals, food and pharmaceuticals	Research on bio-technology including recombinant DNA and cell culture techniques, and their application in pharmaceutical production.
Association for the Progress of New Chemistry	More than 35 chemical firms	Research on high-technology, biotechnology and new materials.

Source: Authors research

FUNDING AND PERSONNEL

MITI's 10-year 'Research and Development Project of Basic Technologies for Future Industries', starting in fiscal year 1981 through 1990, is budgeted for a modest $148.6 million (¥26 billion) with $62.9 million for bioreactor research, $57.1 million for recombinant DNA, and $28.6 million on large-scale mass cell culture research. In fiscal year 1985 (April 1985 through March

1986), the government share of total biotechnology R & D expenditures in Japan, $820.8 million (¥143.6 billion), amounted to about 34 per cent or $277.9 million. Of total government expenditure, MITI's share is only about 8.7 per cent or $24.2 million, while Ministry of Education funding accounted for 83.3 per cent or $231.4 million.

Private sector spending ($542.9 million in fiscal year 1985) accounts for the bulk of total biotech R & D expenditure in Japan. A recent survey among some 268 corporations indicated average per-company biotech R & D expenses at $1.7 million in fiscal year 1985, up 51.5 per cent over the preceeding fiscal year. In fiscal year 1984, private R & D spending averaged $1.1 million per company, up 28.2 per cent over fiscal year 1983. Since fiscal year 1980, annual increases in total private sector biotechnology R & D expenses have hovered around 20 per cent as compared with 10–12 per cent for all R & D expenditures.

TABLE 5.7   BIOTECHNOLOGY R & D FUNDS FISCAL YEAR 1985[1]

	$ million
Government sector:	
Ministry of International Trade and Industry	24.2
Ministry of Health and Welfare	15.4
Ministry of Agriculture, Forestry and Fisheries	6.9
Ministry of Education	231.4
Sub-total	277.9
Private sector:	
Corporations	542.9
Total	820.8

Notes: [1] April 1985 – March 1986
Exchange rate: $1=¥175
Sources: Respective ministry; company records

By industry, annual biotechnology R & D expenditures are increasing by about 30 per cent in the food sector, by some 20 per cent in pharmaceuticals, chemicals, textiles, and pulp and paper, while the rate is about 10 per cent or less in others, such as plant engineering and oil companies.

The rate of biotech R & D spending to total R & D spending in Japan is steadily rising. Based on annual expenditure growth rates in 1980–1984, the ratio was at least 8.0 per cent in fiscal year 1985. The ratio is particularly high in the food industry, about 30 per cent. In chemicals, textiles, and pulp and paper, the ratio is approximately 10 per cent, while the pharmaceutical industry still records a conspicuously low ratio at less than 10 per cent, reflecting the fact that medical products are still mainly developed through chemical synthesis processes rather than through biotechnology.

The number of biotechnology researchers in the private sector is estimated to be about 4000, according to a MITI survey, representing 5.1 per cent of all researchers in the bioindustry. Biotech R & D staff has been increasing by 12 per cent annually in recent years as compared with 3.5 per cent for general research staff, underscoring the highly labour-intensive nature of biotechnology research. The ratio of biotechnology to general researchers is particularly high in the food industry, about 25.5 per cent. In pharmaceuticals, chemicals, textiles, and pulp and paper, the ratio is well below 10 per cent. However, the rate of increase of biotechnology R & D staff is about 20 per cent annually in textiles, pulp and paper. A similar growth trend is noticeable in pharmaceuticals and foods.

The acquisition and cultivation of competent R & D staff are of crucial importance to Japanese corporations at this stage. It is generally known that the Japanese bioindustry has been suffering from a lack of properly educated academic scientists, and it will be recruiting overseas to overcome this handicap. It is also quite possible that more Japanese corporations will set up laboratories overseas, staffed by native researchers.

MAJOR PROJECTS

MITI's 'Research and Development Project on Basic Technologies for Future Industries' singles out three of the above fields for intensive research:
Recombinant DNA, targeting the development of technologies for new microbe construction yielding chemical substances for industrial uses. Four areas are highlighted: Basic technology of recombinant DNA; Construction of micro-organisms for high oxidation reactions; Construction of an efficient secretory host-vector system in B subtilis; Development of secretory host-vector systems in yeast;

91

Large-scale cell culture, aiming at developing a serum-free culture medium and technology to cultivate cells in on a large scale and in high density.

Bioreactors, aiming at the production of catalysts providing an alternate, lower energy pathway for chemical reactions. Research on potential applications falls into four areas: Immobilisation of enzymes; Improvement of enzyme functions; Recycling co-enzymes; Substitute catalysts.

These technologies are already being used in a broad range of industrial fields including the pharmaceutical industry for producing growth hormone, insulin, interferons, and monoclonal antibodies; the food industry for amino acids and development of new breeds of plants; and the chemical industry for resource- and energy-conserving processes by using a bioreactor and for resources and energy by using biomass (see Table 5.9). R & D trends by product lines indicate the majority of R & D efforts is being focused on growth hormones, insulin, thromboclasis substance and albumin; on blood serum by using the gene recombinant technology; and on fine chemicals with high added value such as interferon and urokinase by applying the cell culture technology.

TABLE 5.8   CURRENT BASIC RESEARCH BY COMPANIES INVOLVED

Fields of research	Number of Companies
Bioutilisation technologies:	
Microbe & cell screening	100
Microbe & cell mutation	85
Recombinant DNA	75
Cell fusion	84
Fermentation	110
Enzyme utilisation	114
Semi synthesis	55
Bioreactor	70
Large-scale cell cultivation (animal)	34
Large-scale cell cultivation (plant)	31
Product separation & refining	92
Plant engineering	64
Others	19
Biomimetic technologies:	
Artificial enzyme-related	8
Biocompatible materials	22
Others	10

Source: MITI Survey, 1982 (157 firms, duplicate replies).

TABLE 5.9 CURRENT APPLIED RESEARCH BY COMPANIES INVOLVED[1]

Fields of research	Number of companies
Chemical products:	
Commodity chemicals	12
Enzyme	82
Colouring matter, pigments	13
Organic acids	39
Perfumes	15
Amino acids	36
Surface active agents	8
Other	24
Medical drugs:	
Physiological active substances	94
Antibodies	49
Antibiotic substances	44
Vaccines	9
Vitamins	26
Other	35
Agricultural chemicals:	33
Feedstuffs and foods:	
Feedstuff	41
Food	48
Food additives	48
Energy – natural resources:	
Methane gas	20
Hydrogen gas	7
Fuel alcohol	27
Other	7
Mining:	
Bacteria leaching	5
Environmental preservation:	
Waste water treatment	61
Other	7
Agricultural products:	17
Others:	
Biosensors	25
Therapy, diagnosis	63
Other	8

Source: MITI Survey, 1982

According to the major 1982 MITI survey on biotech R & D in the private sector, covering some 157 corporations, over 60 per cent of these were conducting research on enzyme utilisation, fermentation, and microbe/cell screening technologies, while over

50 per cent were pursuing microbe/cell mutation technology research. About half of corporations surveyed were engaged in 'new' biotechnology research, such as recombinant DNA, cell fusion and bioreactors; while 20 per cent were researching technologies for animal and plant cell cultivation.

TABLE 5.10   KEY AREAS OF FUTURE RESEARCH BY
COMPANIES INVOLVED

Fields of research	Number of companies
Bioutilisation technologies:	
Chemical products	83
Medical drugs	103
Agricultural chemicals	29
Feedstuff, food	59
Energy, natural resources	33
Mining	2
Environmental preservation	27
Agriculture	20
Others	60
Biomimetic technologies:	
Artificial enzymes	2
Biocompatible materials	9
Others	5

Source: MITI Survey, 1982.

Pharmaceutical products have probably received the most publicity and the Ministry of Health and Welfare has already put recombinant DNA human insulin on its official drug list, and is shortly expected to approve human growth hormone and alpha-interferon, the first recombinant pharmaceutical product to be actually manufactured in Japan. Clinical trials are already being conducted on interleukin-2, hepatitis B vaccine, tumour necrosis factor (TNF), beta- and gamma-interferon and tissue plasminogen activator (TPA).

Commercialisation is proceeding very rapidly on a great variety of diagnostic tests capable of high precision detection and diagnosis of viruses, bacteria and cancers. Some of these involve 'in vitro' imaging where labelled antibodies migrate to the diseased area and can be visualised on a screen. Most are based on laboratory analysis of blood samples. These 'in vitro' tests are often just improvements of assays which have been on sale using polyclonal antibodies, but can now be made more sensitive, specific and reliable using monoclonals.

Production costs are a basic issue in the chemical and food industries, and the current focus of Japanese R & D is on bioractors to lower production costs, increase production volumes and improve quality. The most promising results involve fine chemicals like amino and nucleic acids. Successful industrial applications through immobilised enzymes have already been achieved. Genetic engineering and cell fusion are still the main technologies used to develop new organisms, however, to pave the way for accelerated commercialisation of genetic engineering in fine chemicals, MITI has established safety guidelines covering both the bacteria used and the steps of production processes. Categories for genetic engineering projects are defined by the degree of safety of genes and host micro-organisms used, called good industry large-scale practice (GILSP), Categories I–III: Category III allows the use of pathogenic host microorganisms, requiring air-tight equipment and fumigation of production facilities. The guideline is expected to apply to food additives, amino acids and industrial-use alcohol initially. Biotechnological processes are estimated to improve efficiency of amino acid production by 20–30 per cent. The Ministry of Agriculture, Forestry and Fisheries is also developing a guideline covering food produced by biotechnology.

Fine chemical applications already being exploited include the amino acid L-tryptophan, bacteria capable of mass producing herbicides, and the cytochrome P450 gene producing the oxidation enzyme in yeast to develop a variety of aromatic compounds. Surfactants and cosmetics are other interesting fields. Bio-cosmetics are mostly manufactured through fermentation and tissue culture. Lipsticks and liquid cosmetics are typical products.

Biotechnology has become crucially important to the food industry with changing dietary habits of consumers toward low-calorie foods. Fermentation and bioreactors are the key technologies for meeting these needs. Japan's centuries-old tradition of fermentation techniques is now being linked to bioreactor technology to shorten processing time. Recent applications of this combined technology are the production of new sweeteners, notably isomerised sugar, a mixture of glucose and fructose. Another application field now under research is single-cell proteins (SCPs), a form of micro-organisms used as protein source in livestock feed. Test marketing of ginseng mass

TABLE 5.11   DEVELOPMENT OF NEW DRUGS THROUGH BIOTECHNOLOGY

Company	Product	Technique	Technology; R & D Target, Development stage
Sankyo Co.	Antibiotics Tissue plasminogen activator Monoclonal antibody	Gene manipulation Cell fusion	Original Licensed from Celltech Original; diagnostic drug
Takeda Chemical Industries	Interleukin 2 Human immuno-globulin IgE Monoclonal antibody	Gene manipulation Gene manipulation Cell fusion	Original; immunity control substances Original; diagnosis and treatment of allergic diseases Original; diagnostic drugs
Yamanouchi Pharmaceutical Co.	Tissue plasminogen activator	Gene manipulation	Licensed from Genex
Dainippon Pharmaceutical Co.	TNF	Cell culture	Tied up with Asahi Chemical Industry
Shionogi & Co.	Human serum albumin Interleukin 2 Human insulin	Gene manipulation Gene manipulation	Licensed from Biogen Licensed from Biogen Licensed from Eli Lilly Co.; Phase III application filed
Tanabe Seiyaku Co.	Amino acids Alcohols	Immobilised micro-organisms Bioreactor	Original; Commercialisation has started. Original
Yoshitomi Pharmaceutical I[dustries	Synthesised DNA		Sales tie up with Yuki Gosei Kogyo Co.

5.11 contd.

Company	Product	Technique	Technology; R & D target, development stage
Fujisawa Pharmaceutical Co.	Tissue plasminogen activator (TPA)	Gene manipulation	Licensed from Genentech
	Physiologically active agents	Gene manipulation	Original
	Monoclonal antibody	Cell fusion	Original; clinical test drugs
Wakamoto Pharmaceutical Co.	Physiologically active agents	Gene manipulation	Original
Banyu Pharmaceutical Co.	Antibiotics	Gene manipulation and cell fusion	Original
Chugai Pharmaceutical Co.	CSF	Cell culture	Original; diagnostic drugs
	Monoclonal antibody	Cell fusion	Licensed from University of South Carolina; diagnostic drugs
	Physiologically active agents	Gene manipulation	Licensed from Genex
The Green Cross Corporation	Colony stimulating factor (CFS)	Cell culture	Joint development with Morinaga Milk Industry
	Urokinase	Gene manipulation	Original
	Hepatitis B vaccine	Gene manipulation	Licensed from Biogen
	Human serum albumin	Gene manipulation	Licensed from Genex
	Monoclonal antibody	Cell fusion	Licensed from UCLA, Tohoku University, Nippon University, Hokkaido University (treatment drugs)
Eisai Co.	Physiologically active agents	Gene manipulation	Original
Mochida Pharmaceutical Co.	Carano-breaking factor (CBF)	Intracellular culture	Licensed by Hayashihara Biochemical Laboratories

Source: The Long-Term Credit Bank of Japan (LTCB)

TABLE 5.12 BIOTECHNOLOGY R & D IN AGRICULTURE

Technique	Company or Institute	R & D activity
Recombinant DNA	(1) Study group at the National Institute of Genetics	Breeding of a nitrogen fixing rice plant. (The tasks ahead are to increase the nitrogen fixing capacity of symbiotic leguminous bacteria by about 10 times and to improve the Japanese native rice variety (Japonica) so that it can symbiose with leguminous bacteria.)
	(2) Study group at the Faculty of Agriculture, Tokyo University	The group cut the DNA producing rennin, an enzyme required in making cheese from milk, from the cell of the fourth stomach of a cow and inserted it in *Escherichia coli*, and has established a process for mass production of rennin.
Cell fusion	(1) Ajinomoto Co.	The company has succeeded in increasing the capacity of the amino acid-producing yeast by about 3 times through cell fusion. (The company has brought into being a bacterium that produces lysine and threonine of high purity, which are used as nutritive additives for livestock feed.)
	(2) Takeda Chemical Industries	The company improved through cell fusion the bacterium that produces enramycin, an antibiotic added to livestock feed, and has obtained a bacterium that produces the antibiotic of a purity about five times as high as achieved by conventional methods.
	(3) Meiji Seika Kaisha	The company obtained a new strain of hypha through cell fusion of ray fungus and has succeeded in the production of a herbicide by use of the new strain of hypha.
Tissue culture	(1) Nippon Paint Co. and study group at the Faculty of Agriculture, Kyoto University	The joint study group has succeeded in mass extraction of a red pigment of high purity called anthocyan through tissue culture of a foliage plant called Hanakirin. (The extraction of this pigment which took 20 years can now be accomplished in no more than 20 weeks' time.)

5.12 contd.

Technique	Company or Institute	R & D Activity
Tissue culture	(2) Dai-ichi Seed Company	The company has succeeded in clonal multiplication of 20 kinds of flowering plants, such as carnation, lily and freesia. (The company has developed a technique with which to multiplicate flowering plants without resorting to fertilisation.)
	(3) Kyowa Hakko Kogyo Co.	The company has succeeded in clonal multiplication of gold-banded lily.
	(4) Nitto Electric Industrial Co.	The company has developed a process for mass production of ginseng through tissue culture.
Embryonic transplantation	(1) Institutes	Studies are under way to realise embryonic transplantation in pigs, sheep and other domestic animals excluding cows. This technique consists of planting several embryos in a cow with a view to obtaining more than a twin after each gestation.
Bioreactor	(1) Ajinomoto Co. (in technical tie-up with C D Sale)	C D Sale has developed a new process for production of a low-calorie amino acid sweetening agent (Aspartame) with the aid of a micro-organism. This process is a break with conventional methods of chemical analysis. Ajinomoto Co. is conducting an experiment in a technical tie-up with C D Sale to commercialise this process.
	(2) Meiji Seika Kaisha	(a) The company has succeeded in production of a herbicide (hypha) by use of a new strain of ray fungus. (b) The company has commercialised a process for planting live *Lactobacillus bifidus*, a digestion-promoting bacterium, in biscuits and other confectioneries to obtain health foods.
	(3) Kikumasamune Jozo Co.	The company has developed a technique to insert a wild yeast killing plasmid (extranuclear chromosome) in wild yeast whereby sake being brewed is not contaminated by wild yeast.

Note: In addition to the companies referred to in the table above, many companies of various industries have been carrying on biotech research and development in the field of agriculture, using a variety of approaches: Kikkoman Shoyu Co. (research of new kinds of plants by use of enzymes developed by the company through cell fusion); Kagome Co. (breeding of resistant varieties of tomato through cell fusion); Kyodo Shiryo Co. (processing of water hyacinth into feed and energy); and Penteru (carboxypeptidase extracted from wheat. The product has already been marketed.)
Source: The Long-Term Credit Bank of Japan

produced with tissue culture technology is scheduled for the near future, and the use of bioreactors to shorten fermentation time for soy sauce, beer and wine has made notable achievements. Research on culturing yeast strains to brew sake (rice wine) and shochu (distilled spirits), and on cold-resistant bread yeast is well advanced. Applications of genetic engineering to foodstuffs are still slow due to concern voiced by the consumer movement.

Although it is expected that biotechnology will eventually have its greatest impact in agriculture, Japanese applications research is not yet too well advanced, reportedly due to limited domestic demand for the new technologies, but also attributable to Japan's self-sufficiency in agriculture products, except rice.

Agricultural applications of biotechnology include:

Recombinant DNA for disease and cold resistance; low-temperature sprouting; and dwarfing. Most research is still in the basic research phase. Flexible plant design is not attainable in the foreseeable future;

Cell fusion for hybrids and crossbreeding. This is still mainly in the basic research phase, but applications are likely before recombinant DNA;

Growth acceleration for improved cultivation under artificial light with artificial nutrients to overcome seasonal limitations;

Tissue culture for virus-free growth, large-scale propagation and cultivation, and plant breeding applications.

The latter two technologies are already well known, and have a wide range of commercial applications. The first two are still in initial experimental stages. Japanese seed companies, the leaders in agricultural biotechnology, maintain high technological levels, but are small in scale. By comparison, biotech R & D in fisheries is well advanced. Salmon and eel growth hormones have been developed by means of genetic engineering with outdoor fish breeding tests scheduled to start in late 1986.

Other application fields of biotech R & D are energy, resources and environmental protection. Ongoing Japanese projects include bacterial leaching for mineral extraction from ores, and treatment of industrial wastes and sewerage by means of bioreactor technology.

5.4 Future Outlook

Being a technology with a long gestation period between research and commercialisation, the real impact of biotechnology will not be felt until after the turn of the century. Various scenarios could be envisaged. Several major biotechnologies will have entered the commercialisation phase. In the medical field, real advances in the understanding of immunological response mechanisms, gene structures, and development of artificial organs will have materialised. Conventional production processes will be replaced by biochemical processes as costs decline, opening up new means of energy generation through biomass conversion. New breeding methods for plants and animals through recombinant DNA should have a favourable impact on agriculture and food production.

The industrial stage of biotechnology will be reached when it is widely used, directly or indirectly, in various other industries. Indications of things to come are already noticeable in foods and pharmaceuticals, but will extend to production of basic industrial raw materials and recycling of wastes. In the later 'social' stage, decoding of human genetic maps, the treatment of hereditary diseases and the artificial reproduction of humans may become possible.

Larger Japanese companies are now considered to have the leading edge in several fields of biotechnology, and are greatly contributing to expanding the range of applications. This is particularly true for fine chemicals, where Japanese companies conduct R & D with their own technology and where there is a real possibility they will contribute significantly to the development of technology worldwide. However, in developing pharmaceutical products through biotechnology, Japanese companies lag seriously behind their foreign counterparts. To overcome this, contracts to obtain technology from firms in Europe and United States have been concluded.

Basic research has traditionally been the weak point of many Japanese companies, who have preferred to concentrate their R & D efforts on applications. Unlike the United States, new venture capital financed businesses have never played a big role in the Japanese bioindustry. Rigid employment practices in larger companies and the structure of domestic capital markets are usually seen as the main reasons for this. These circumstances may well have been the prime motivation of the government, and particularly MITI, to establish various bodies such as the Key Technology Center and others, to provide technical and capital support for small- and medium-sized companies in the bioindustry.

TABLE 5.13   FUTURE IMPACT OF BIOTECHNOLOGY ON INDUSTRY IN JAPAN

General	Scale of economy and industrial structure	The industrialisation of agriculture will accelerate. There will be an increasing fusion of nonbiotechnologies and biotechnology.
	R & D	R & D costs will be reduced. There will be an increasingly short supply of biotech researchers. Active interchange will occur between universities and businesses. Independent institutions will have to be set up to coordinate universities and businesses.
	Patents	Lawsuits concerning biotech patents will increase. Overseas licensing will stagnate Japanese industry.
Resources and energy	Supply	There will be progress in conversion from oil to biomass. Competition will intensify between supplies of foods, resources and energy produced from biomass and those from existing sources. Energy supply will become unstable owing to abnormal weather. Local energy use will increase.
	Consumption	There will be progress in conservation of energy and resources.
	Transportation	Gasahol cars will become more popular.
Industry	Capital investment	The size of plants will become smaller. There will be a growing need for safeguards in industrial facilities to prevent the escape of living organisms.
	Prices	The prices of fine chemicals will decrease. Food prices will stabilise.
	Market	The market for petroleum chemicals will shrink.
	Product quality	Drug side-effects will decrease. Chemical products' raw materials will diversify.
	New products	The development of new drugs will advance. Designed compound products will increase.
	Labour conditions	Physical and chemical hazards in the workplace will decrease. Workers will be required to gain knowledge of handling living organisms. Labour saving will advance in agriculture.
	Conditions of plant location	There will be a relaxation of regulations on plant location.
	Social costs	Social costs will decrease.
International relations	Exports	Exports of high value added products will increase.
	Imports	Oil imports will decrease. There will be an increase in dependence on imports from biomass-rich countries. Food self-sufficiency will increase.
	External investment	Plant exports to developing countries will increase.

Source: Ministry of International Trade and Industry (MITI), Biotech Industry Office

The Japanese bioindustry is growing fast, but its economic impact is not yet tangible. According to various MITI reports, the path to commercialisation will be led by pharmaceuticals, followed by foods additives and flavourings, paper-pulp products and synthetic rubber. Along the road in the years up to 1990, differences between individual companies will grow markedly, particularly in the pharmaceutical field. Shakeouts are likely to structure the bioindustry along different lines towards more specialisation and concentration in every field of endeavour, but the overall trend is to an industry largely controlled by existing major companies.

6 Chapter Six
The Biotechnology Industry in Western Europe

6.1 Introduction

NATIONAL STRUCTURE

In looking at the biotechnology industry in western Europe and comparing it with the American and Japanese experiences, it is of course necessary to start by saying that Western Europe comprises twenty separate countries. The development of the industry has consequently been piecemeal in nature, very much depending on government and industry attitudes in each country.

THE TWENTY COUNTRIES OF WESTERN EUROPE

EEC	Other
Belgium	Austria
Denmark	Cyprus
Eire	Finland
France	Iceland
Greece	Malta
Italy	Norway
Luxembourg	Sweden
Netherlands	Switzerland
Portugal	
Spain	
United Kingdom	
West Germany	

Of these twenty countries, the most substantial progress toward commercialisation of biotechnology products has been made by West Germany, followed by the United Kingdom and then France. However, the Swiss company Biogen has been at the cutting edge of basic research in biotechnology; a Dutch company was one of the first to produce a commercial product; and Scandinavian companies have successfully used their specialist knowledge (in hormones, insulin and enzymes) to push back the frontiers of R & D in biotechnology. These contributions should not therefore be ignored, as no country has a monopoly on genius or dedication in the laboratory.

The various countries are taking different paths in their approach to biotechnology development. France has the most comprehensive, forward-looking national plan, whereas in the UK development has been more fragmented, throwing up many small venture-capital companies. The German path has been cleared by the massive chemical concerns headquartered there, together with ample funds provided by non-profit organisations. In Switzerland, Biogen – for many years the undisputed leader among European biotechnology specialists – was the product of American as well as Swiss private investment. Cross-fertilisation of initiatives in this field have been characteristic of the industry since its inception in the late 1970s. This has involved more significant joint projects between European and American or Japanese companies than between companies within Europe, suggesting a degree of inter-nation jealousy within Europe as well as a need to seek out global markets.

The European Commission (EC) will play a primary role in the development of biotechnology research and exploitation in the ten countries of the European Economic Community (EEC). However, in the final analysis the development of the industry will still depend on the degree of commitment of the member nations. So far, the major contributors to development in the EEC have been West Germany, France, and the United Kingdom. The EC has encouraged a more united approach to biotechnological development in the Community, but by its very bureaucratic nature progress has been slow, lagging behind the USA and Japan. However, it has been argued that in the long run this will provide a more stable, if slower, path of development. At the same time, individual countries and national companies are free to develop their own, small-scale projects, and to react quickly in the laboratory if not in the commercial market to new discoveries.

COMPANIES IN BIOTECHNOLOGY

Europe boasts only a small number of specialised biotechnology companies. The discovery of the potential for biotechnological products coincided with the worldwide recession, and this made venture capital hard to come by in most European countries. The outstanding specialist companies in Europe by the mid 1980s were Biogen (Switzerland), Celltech (UK), and Genetica and Transgene (France); German research is centred on chemical companies. However, a number of companies have come to feature strongly by already being 'pre-biotechnology' specialists in related fields where exploitation of new discoveries is ongoing. Good examples are Gist-Brocades (Netherlands), a large producer of enzymes and crude penicillin and an expert in fermentation technology, and in Denmark, Novo Industri and Nordisk Gentofte, Europe's largest producers of insulin.

In Europe, as in the USA and Japan, the pharmaceutical applications of biotechnology are considered to be the most relevant in the short- to medium-term future of the industry. American multinationals are already prominent in the European pharmaceutical markets, and often dominate certain types of chemotherapy. However, Europe's largest drug companies are transnational in the fullest sense, often making most of their profits in overseas markets (including the USA). The largest of these are Bayer, Hoechst, Schering and Boehringer Ingelheim (West Germany), Imperial Chemical Industries, Glaxo, Beecham and Wellcome (UK), Ciba-Geigy, Hoffmann-La Roche and Sandoz (Switzerland).

Europe has no shortage of major companies in the general fields of food, chemicals, agriculture and energy which will ultimately have the widest resources to exploit biotechnology. West Germany is particularly strong, with Bayer, Hoechst and BASF ranking among the world's top ten chemical concerns. Imperial Chemical Industries (ICI) of the UK and Anglo-Dutch Unilever are of major importance. Otherwise, individual national industries tend to be dominated by one or two large chemical multinationals — Montedison (Italy), Akzo (Netherlands), Ciba-Geigy (Switzerland), Rhône-Poulenc (France) and Solvay (Belgium). Royal Dutch Shell (UK-Netherlands) is Europe's largest oil/chemicals company. Other large oil companies are usually state owned or state controlled. The largest food/consumer product multinationals based in Europe are Nestlé (Switzerland) and Unilever (UK-Netherlands).

106

PROSPECTS

Europe offers both advantages and disadvantages in the growth of biotechnology. A major advantage is the strength of academic research in Europe, although it must immediately be countered that industry is relatively slow to exploit academic research commercially. Also, there is a constant 'brain drain' of biotechnology experts to the USA which depletes the European potential: this is an abiding concern of the European Commission. A second advantage is the international strength in depth of the largest European chemical companies. The process of colonization was often responsible for the rise of these companies, and they have benefited by coming into contact with and working within tropical and sub-tropical environments with their varying and diverse biological ecospheres.

A competitive spirit between nations and, even more so, between the major European multinationals has been productive in the past, but can now be construed as a disadvantage. The development of biotechnological products and techniques is likely to be advanced by cooperation rather than competition; European companies have become used to 'internal' battles rather than concentrating on a united front against the expansion of American and Japanese R & D. The political environment is usually disadvantageous in Europe. Governments in most countries tend to swing like a pendulum from right to left and back again, and this creates uncertainty for companies which need to plan their R & D and commercialisation projects over ten or twenty years. France has provided a classic example of this, with several major industrial and financial companies having been nationalised and then privatised as governments changed. Meanwhile, even right-wing governments are insisting on cutting the profits of drug companies by reducing the amount paid for medicines by the government health agencies. This is persuading many companies to turn to less research-intensive markets (e.g. household drugs, toiletries), and ultimately acts as a disincentive to research in biotechnology.

It is the task of the EEC to unite Community countries and the companies they host in order to challenge the dominance of the USA and Japan. So far, however, individual initiatives from companies like Biogen and Celltech have been most outstanding. In the long run, a handful of companies could well pose the only threat to American and Japanese dominance of the world biotechnology industry.

6.2 Structure of the European Industry

Europe comprises twenty countries of great variation in size and industrial development. The four largest countries account for 65 per cent of Europe's 355 million inhabitants and nearly 70 per cent of the Gross Domestic Product of the region. The GDP per-capita measure is higher for the top three countries than for the region as a whole.

TABLE 6.1   EUROPE'S FOUR LARGEST MARKETS, 1985

	Population (mn)	GDP $bn	GDP per capita ($)
West Germany	61.0	625	10250
France	54.9	510	9300
United Kingdom	56.6	450	8000
Italy	57.2	360	6300
Other Europe[1]	125.5	875	6970
Total Europe	355.0	2820	7945

Note:[1] 16 countries; see list in Table 1
Source: National statistics

The industries serving the large and numerically similar populations of the four major countries naturally benefit from a massive demand for industrial and consumer goods. In the long run, this bodes well for their development of biotechnological products for consumption or industrial use. However, industrial development in this field has been unequal.

Italy's biotechnology industry lags behind that of West Germany by a long way, for example, and this can be directly related to per-capita wealth. Thus it follows that several of the other smaller European countries which boast per-capita income above the regional average are contributing more to biotechnological developments. Notable in this way are Switzerland, Denmark and Sweden. Switzerland in particular has a crucial significance as the home of a stable economy, high standards of living, traditionally strong technical research, and several large research-based companies which are out of proportion with the Swiss market itself – Ciba-Geigy, Hoffmann-La Roche, and Sandoz.

GENERAL INDUSTRIAL COMPANIES

Broad-based chemical companies are most likely to reap the benefits of biotechnology in the long term. In Europe, many of the largest industrial companies are involved directly or indirectly in many relevant areas – chemicals, agriculture, pharmaceuticals and energy. As nationally dominant enterprises, they have not had to specialise in particular markets as much as most American companies.

This characteristic means that the following list of Europe's largest chemical companies acts as a fairly comprehensive guide, not only to the region's largest chemical concerns but to most of the major companies in biotechnology-related work. The list also includes the chemical subsidiaries of Europe's largest energy companies, plus a sprinkling of Europe's largest consumer-product multinationals (Unilever, Nestlé, Beecham), all of which qualify for inclusion because of large chemical divisions. The forty companies range considerably in size, both in terms of total assets – ICI of the UK is easily the richest company – and in terms of dollar sales. On the latter measure, the top four companies were well clear of the field in 1985. Thirty-six of the companies had sales over $1000 million. The following features may also be noted:
— West Germany is dominant, with ten of the forty and the top three rankings for Bayer, Hoechst and BASF. The UK was second with eight companies (excluding the Anglo-Dutch companies Unilever and Shell).
— Sixteen companies are internationally significant producers of pharmaceuticals, led in sales of these products by Hoechst, Bayer, Ciba-Geigy, Hoffmann-La Roche, Sandoz, Rhône-Poulenc and ICI. Six are chemical subsidiaries of oil companies, of which Shell is the largest. The remainder are principally known for bulk chemical and synthetic products, with the notable exceptions of Unilever (foods, cleaning products) and Courtaulds (textiles).
— Five American subsidiaries in Europe have been excluded. These companies, with their regional sales in 1985, were: Dow Chemical Europe/Africa/Middle East ($4072 million), Essochem Europe ($2700 million), Monsanto Europe/Africa ($1152 million), Du Pont Europe/Africa/Middle East ($1033 million) and Union Carbide ($871 million). If included above, they would have ranked respectively in 13th, 25th, 35th, 39th and 41st positions.

**TABLE 6.2   EUROPE'S FORTY LARGEST CHEMICAL
COMPANIES IN 1985**
($m, ranked by sales of chemicals)

Company	Country	Total assets	Sales	R & D	R & D as % of sales
Bayer	West Germany	3103	15117	687	4.5
Hoechst	West Germany	9901	10564	639	4.4
BASF	West Germany	8512	14192	436	3.1
ICI	UK	11872	13182	398	3.0
Shell	UK/Neths	750	8296	172	2.1
Ciba-Geigy	Switzerland	9563	7435	620	8.3
Unilever*	UK/Neths	n.a.	6187	322	5.2
Rhône-Poulenc	France	4957	5861	282	4.8
Elf Aquitaine*	France	2997	5487	177	3.2
Neste Oy	Finland	2651	5363	14	0.3
Akzo	Netherlands	3562	5147	183	3.6
Montedison*	Italy	n.a.	4986	—	n.a.
Solvay	Belgium	3212	3877	127	3.3
Hoffmann-La Roche	Switzerland	5562	3691	510	13.8
DSM*	Netherlands	1582	3316	118	3.6
Henkel	West Germany	969	3282	76	2.3
Sandoz	Switzerland	3229	3163	270	8.5
Enichem	Italy	1000	3137	65	1.9
Courtaulds	UK	1876	3057	n.a.	n.a.
BP Chemicals	UK	914	2973	38	1.3
CdF Chimie	France	1728	2939	n.a.	n.a.
Beecham	UK	1973	2916	106	3.6
BOC	UK	3892	2874	60	2.1
Atochem	France	1760	2477	n.a.	n.a.
Norsk Hydro*	Norway	n.a.	2372	n.a.	n.a.
Huls	West Germany	1058	2200	56	2.5
Glaxo	UK	1665	1752	126	7.2
Schering	West Germany	1549	1717	163	9.5
Boehringer-Ingelheim	West Germany	1568	1449	197	13.6
Dynamit Nobel	West Germany	225	1272	30	2.4
Roussel-Uclaf	France	1032	1244	108	8.5
Snia BPD	Italy	1497	1191	43	3.6
Wellcome	UK	1138	1147	137	12.0
Rutgerswerke	West Germany	528	1144	n.a.	n.a.
E. Merck	West Germany	488	1060	64	6.0
Superfos	Denmark	464	880	5	0.5
Nobel	Sweden	1511	851	n.a.	n.a.
Chemie Linz	Austria	445	838	25	3.0
RTZ*	UK	588	788	n.a.	n.a.

Note: * chemical sales only
Source: *European Chemical News*

R & D expenditure ranges from a few million dollars to over $500 million in the case of four companies – Bayer, Hoechst, Ciba-Geigy and Hoffmann-La Roche. As a proportion of sales, R & D usually falls between 2 per cent and 5 per cent, but can be much lower or much higher. This depends on the nature of the company's principal business, but also on how R & D expenditure is defined for the purposes of company accounts. Highest proportionate expenditure relative to sales was achieved by pharmaceutical producers: Hoffmann-La Roche (13.8 per cent), Boehringer Ingelheim (13.6 per cent), Wellcome (12.0 per cent), Schering (9.5 per cent), Sandoz (8.5 per cent), Roussel-Uclaf (8.5 per cent), and Ciba-Geigy (8.3 per cent). The presence of the three largest Swiss companies is significant in this list.

Within the area of chemical research and production, pharmaceutical companies are usually at the forefront in biotechnological research. The following table isolates the main pharmaceutical companies in Europe by country, details of their involvement following in a later section. In some cases, pharmaceuticals account for less than a quarter of group sales, but the overall size of the company is enough to ensure national leadership for the group, ahead of many specialist companies. This is most apparent in the cases of Bayer and Hoechst (West Germany) and ICI (UK).

TABLE 6.3   EUROPE'S LEADING NATIONAL PHARMACEUTICAL GROUPS, 1985

West Germany	Hoechst	Boehringer Ingelheim
	Bayer	Boehringer Mannheim
	Schering	E. Merck
United Kingdom	Glaxo	Beecham
	ICI	Wellcome Foundation
	Boots	
Switzerland	Ciba-Geigy	Hoffmann-La Roche
	Sandoz	
France	Rhône-Poulenc	Roussel-Uclaf*
	Sanofi	
Italy	Montedison (owns Farmitalia Carlo Erba)	
Netherlands	Akzo	Gist-Brocades
Sweden	Astra	Kabi Vitrum
Denmark	Novo Industri	Nordisk Gentofte

Note: * Roussel-Uclaf partly owned by Hoechst (West Germany)
Source: Authors research

In addition to chemical companies, biotechnology is proving of interest to companies in various other industrial sectors in Europe – food, alcoholic drinks, and oil companies being the principal sectors. However, relatively few of these are multinational in scale, since even oil companies are frequently state owned and are therefore primarily concerned with domestic matters. Other major oil companies are subsidiaries of American corporations (Esso, Texaco, Mobil), with USA headquarters deciding corporate policy and planning on such long-term issues as investment in biotechnology research.

TABLE 6.4 EUROPE'S LARGEST OIL/ENERGY COMPANIES IN 1985 ($ billion)

Company	Country of parent	Sales
British Petroleum	UK	62.9
Royal Dutch Shell	Netherlands	58.5
Shell Transport/Trading	UK	39.0
ENI*	Italy	26.9
Elf Aquitaine*	France	22.4
Total France	USA	19.7
Esso UK	USA	13.5
Petrofina	Belgium	13.0
Shell UK	UK	11.7
Esso West Germany	USA	9.0
Deutsche BP	UK	8.3
Statoil*	Norway	5.6
Deutsche Texaco	USA	4.7
Texaco UK	USA	4.6
Esso France	USA	4.4
Esso Italy	USA	4.4
Empresa Nac. del Petroleo*	Spain	4.3

Note: * partially or totally state owned
Source: Company research

Very few companies in the agriculture, food and drink industries have the resources to find substantial R & D funds for this type of research. The main burden of research into ways of improving yields and food output is carried by Europe's major chemical companies, as the food and drink industries are usually much more fragmented than in the USA. The following is a list of major industrial companies of European origin which are known to have made investments in biotechnology research.

TABLE 6.5 PRINCIPAL NON-CHEMICAL INDUSTRIAL COMPANIES INVESTING IN BIOTECHNOLOGY, 1980–1985

Industry	Company	Country of origin
Food industry	Unilever	UK/Neths
	BSN Gervais Danone	France
	Rank Hovis MacDougall	UK
	Swedish Sugar	Sweden
	Tate and Lyle	UK
Alcohol/brewing	Carlsberg	Denmark
	Heineken	Netherlands
	Stella Artois	Belgium
	Moet Hennessey	France
Oil companies	Shell	UK/Neths
	Elf Aquitaine	France
	Petrofin	Netherlands
Financial etc.	Banque Paribas	France
	Volkswagenwerk Foundation	West Germany
	Thyssen Foundation	West Germany
	B & C Shipping	UK
	Midland Bank	UK
	DSM (Dutch State Cos.)	

Source: Authors survey

To complete the picture of European industry, the following table lists the largest industrial companies (excluding oil and chemical concerns) by sales, all of which have substantial funds at their disposal but which have only a partial interest in the ultimate products of biotechnology (i.e. food, brewing, distillation of alcohol, agriculture).

SPECIALIST BIOTECHNOLOGY COMPANIES

As in the USA and Japan, none of the emergent specialist companies in European biotechnology has yet generated sales of anywhere near the major chemical, food or energy companies. However, they are only just beginning the conversion from being purely 'ivory tower' companies to commercial companies in the mid 1980s. In the development of biotechnological research,

TABLE 6.6   LARGEST INDUSTRIAL COMPANIES OTHER THAN OIL/CHEMICALS
(whether or not investing in biotechnology)

Company	Country	Main products	Sales ($b)
Nestlé	Switzerland	Food, dairy	20.5
BAT Industries	UK	Tobacco	16.7
Unilever	Neth.	Food, household	14.3
INI (a)	Spain	Holding company	10.5
Unilever	UK	Food, household	8.7
Grand Metropolitan	UK	Food, drinks	7.5
Imperial Group (b)	UK	Tobacco, food	6.6
Allied-Lyons	UK	Food, drinks	4.4
Hanson Trust (b)	UK	Tobacco, food	3.6
BSN Gervais Danone	France	Food, drink	3.4
Bass	UK	Drinks	3.2
Lonrho	UK	Holding company	2.6
Boots Company	UK	Drugs, cosmetics	2.8
Cadbury Schweppes	UK	Soft drinks, food	2.5
United Biscuits	UK	Bakery	2.4
Heineken	Neth.	Brewing	2.3
Tate and Lyle	UK	Sugar	2.2
BayWa	W. Germany	Agriculture/horticulture	2.2
Whitbread	UK	Alcoholic drinks	2.0
RHM Foods	UK	Food, bakery	1.8
Distillers Co. (c)	UK	Alcoholic spirits	1.7
Reckitt & Colman	UK	Household, food	1.7
EMC (d)	France	Fertiliser, feeds	1.7
Booker	UK	Poultry, foods	1.6
Arthur Guinness (c)	UK	Beer, spirits	1.6
Beghin-Say	France	Sugar, paper	1.5
Union Laitiers	France	Dairy	1.5
Wessanen	Neth.	Food	1.5

Notes: (a) INI is a Spanish state-investment group
      (b) Imperial bought by Hanson in 1986
      (c) Distillers bought by A. Guinness in 1986
      (d) Entreprise Minière et Chimique (mining, chemicals)
Source: Company research

Europe has played an important role, with several major breakthroughs. The most celebrated achievements on this side of the Atlantic took place in Cambridge, England, in 1975, when Millstein and Koehler succeeded in producing the first monoclonal antibodies. Credit for this breakthrough in MAB research goes to the British government-funded Medical Council. However, the UK provides a relatively supportive background for biotechnological research. Several companies, like Beecham and Glaxo, are long-time specialists in microbial research as well as being international producers of anti-infective drugs. The Wellcome Foundation is probably the world's leading commercial authority on anti-virals and is a strong contributor in other biological fields of research.

The first of Europe's major specialists to be formed, however, was Biogen in Switzerland in 1978. The company was in fact initiated by Americans – Harvard University, backed by funds from Schering-Plough, INCO and Monsanto – and soon came to rank in academic standing with the other major USA-funded companies, Genentech, Cetus, and Genex. A total of $3 billion was invested in the company between 1978 and 1986. Schering-Plough and Monsanto subsequently sold their Biogen shares (in 1986) probably in order to concentrate on their own in-house research. Biogen is significant for the breadth of its research, which ranges from interferon and insulin, the earliest biotechnological products, to vaccines, growth hormones and diagnostic kits. The company has more significant contracts with foreign commercial companies than any other in Europe, and has specialised in deals with leading Japanese companies which will allow them to use rDNA techniques developed by Biogen in industrial production of various biotechnological products.

TABLE 6.7   BIOGEN – INTERNATIONAL AGREEMENTS, 1980–1986

Company	Research	Contract/Agreement
Green Cross (Japan)	Hepatitis-B vaccine	Green Cross to produce, market
Shionogi (Japan)	Gamma-interferon	Joint R & D; Shionogi to develop fermentation techniques
Teijin	Blood-clots	Joint development
Meiji Seika (Japan)	Agricultural chemicals	Meiji Seika to use Biogen's rDNA techniques

Source: Company information

Biogen has its own American subsidiary, Biogen Incorporated, in Cambridge, Massachusetts, which will produce biotechnological products in-house, as well as licensing the company's technique to other commercial producers. Unfortunately, later developments have not been kind to Biogen. Its plant on alpha-interferon has been rescinded because the specifications in the patent were decided to be too broad. In 1986, the company decided to rationalise its operations in two ways – by selling off its Swiss research operation, and by concentrating on fewer applications. The main focus now is on developing an injectable gamma-interferon for cancer and arthritis sufferers. Commercialisation is hoped to start in 1987 or 1988 in Europe and by the end of the decade in the USA.

If Biogen has been struggling in recent years, then the UK's main specialist, Celltech, has been consolidating its position in European biotechnology. Celltech was set up in 1980 with 44 per cent of its funding coming from the British government's Enterprise Board (now defunct). The remaining shares were equally divided (14 per cent each) between British and Common-wealth Shipping, the Midland Bank, Prudential Assurance, and Investors in Industry (a venture-capital consortium in the City of London). Like Biogen, Celltech has experienced shifts in owner-ship. The British Technology Group – successor to the Enterprise Board – sold its remaining 15 per cent stake in 1986, and B & C Shipping became the largest shareholder. Shares are also held by Rothschilds' Biotechnology Investments. Celltech's main commer-cialisation agreement has been with The Boots Company, a leading British retail and pharmaceutical company. Boots-Celltech is developing diagnostic kits for fertility and 'flu diagnosis, using Celltech's research into monoclonal antibodies. Another major agreement, in 1985, was with Johnson & Johnson (USA) to produce commercial quantities of MABs.

In 1986, Celltech announced a turnover of $15 million, double the 1985 figure; a $1 million loss in 1986 is forecast to turn around into the company's first profit in 1987. Celltech is involved in several sectors of biotechnological research – rDNA, MABs including interferon and blood reagents, hybridomas, diagnostics – but the company's main sales come from monoclonals, of which it sold some $5 million worth (3Kg) in 1986. This gave Celltech a world-leading 2 per cent share of the international market for MABs.

Biogen and Celltech have become the most outstanding

**TABLE 6.8   CELLTECH: INTERNATIONAL AGREEMENTS
1980–1986**

Company	Research	Contract/Agreement
Boots (UK)	Diagnostics	Boots-Celltech handles sales
Johnson and Johnson (USA)	MABs	J & J to produce
Shell (UK)	Various	Product development
Agrigenetics (USA)	Hybridomas	Product development
Sumitumo (Japan)	All areas	Sumitumo has 5-year contract to act as exclusive agent for Celltech in Japan
Sankyo (Japan)	TPA, MAF	Product development
Serono Laboratories	HGH	Product development

Source: Company information

European biotechnology specialists. However, the trend since 1984 has been for large chemical, pharmaceutical and oil multinationals to set up their own research laboratories in-house, rather than investing in R & D specialists. This has always been the structure of the biotechnology industry in West Germany (see West Germany profile below), where biotechnology research is dominated by the massive chemical concerns – Bayer, Hoechst and the Boehringer companies. The following table therefore indicates only those prominent 'independent' specialists, and not the in-house departments of multinationals where both research and industrial development are increasingly taking place. In many cases, of course, larger companies have invested funds (up to subsidiary level) in these 'laboratory' enterprises, and these are indicated by an asterisk.

6.3   R & D: Investment and Organisation in European Biotechnology

Accurate assessment of the level of interest in biotechnology in Europe in terms of hard cash investments is a virtually impossible task because of the complex cross-fertilisation arrangements between European, American and Japanese companies and because of the number of different countries involved. Each country channels government funds into biotechnology to a lesser or greater degree, but often these funds are not centralised, being

TABLE 6.9 PRINCIPAL SPECIALIST BIOTECHNOLOGY FIRMS, 1985

Belgium	Biotec SA
	Plant Genetics Systems
	UCB-Bioproducts*
Denmark	Novo Industri
	Nordisk Gentofte
France	Genetica*
	Transgene
	Immunotech
	Elf-Bioindustrie*
	Institut Merieux
	Institut Pasteur
West Germany	(see text above)
Italy	Biotec, Bologna
	Pierrel
	Serono
	Sorin Biomedica
Netherlands	Gist-Brocades
	Intervet*
Spain	Immunologia y Genetica Aplicado
	Lloret
Sweden	Fermenta
	Kabi Gen*
	Fortia Pharmacia
Switzerland	Biogen
	Chemap
	PetroGenetic
	Uniserum
United Kingdom	Celltech
	Amersham International
	Agricultural Genetics
	IQ (Bio)
	Sera-Lab
	Porton International

Note: * subsidiaries of larger companies
Source: Authors survey

passed through various departments or ministries such as Agriculture, Industry or Education and Science. Cash is also provided by the EEC, but one of the European Commission's primary tasks for the next few years is seen to be the need to assess the current investment situation. This will entail coordination of statistics-gathering from well over 300 organisations in Europe. It is estimated that around one third of the world organisations undertaking biotechnology R & D are located in Europe (including Israel).

118

TABLE 6.10   ORGANISATIONS INVOLVED IN BIOTECHNOLOGY R & D, 1983

	Public institutes	Private corporations General	Specialist
Europe + Israel	162	106	32
N. America	137	111	146
Japan and other Australasia	51	83	2
Total	350	300	180

Source: *Biotechnology Newswatch* (McGraw Hill)

The figures above seriously underestimate European involvement in biotechnology in one category – the number of specialist companies set up to exploit or research biotechnology. There are probably nearer 80 specialist companies in Europe. Also, to the number of public research institutes can be added another 20 or so major universities where research is being undertaken into biotechnology, frequently with private as well as public financing.

In order to put a value on European R & D in biotechnology it has been necessary to accumulate the values of grants, investments and research budgets in the major countries in a general way. With the inclusion of EEC budgets, it is estimated that a total of $3300 million has been invested in major projects in the 1980–1986 period in the whole of western Europe. This excludes venture capital, worth at least $10,000 million, and the unquantifiable expenditure incurred on in-house R & D by major multinationals. EEC figures for the 10 Community countries, which to date are available only for all forms of R & D of whatever nature, in large categories such as 'Technology' and 'Defence', indicate that 85 per cent of all R & D in the EEC is accounted for by expenditure in three countries – West Germany, France and the UK. This concerns all forms of research, including university, industrial, agricultural, space exploration, social science and health research. In the specific case of biotechnology, it is accepted that the largest nations account for an even higher proportion of the European total. West Germany, France and the UK account for between 25 per cent and 30 per cent each of spending on biotechnology in the Community. However, a less concentrated pattern emerges when the whole of Europe is considered, since this brings in research conducted in Scandinavia and Switzerland.

CENTRAL ORGANISATIONS

The importance of the European Community is still increasing as more nations join the Community – Spain, Portugal and Greece having been admitted in the 1980s. At the beginning of the decade, the governing Commission allocated $8 million for a four-year research programme in bioengineering, with a focus on fermentation technology and agricultural applications. As this amount is much less than the amount required to set up even a fairly modest plant in the private sector, it can be seen that for the foreseeable future the EEC will not be a major source of funding for any 'European' ventures. The more important role of the EEC will be in coordinating public research activities in the Community to ensure that there is a free interchange of information (on non-commercial projects) and that the overlap in research is kept to a minimum. FAST, the EEC programme on Forecasting and Assessment in Science and Technology, is charged with this responsibility.

The European Federation of Biotechnology is the first specific international organisation in Europe concerned with biotechnology. The location of the Federation headquarters in West Germany is logical since the Germans were the first to create a national organisation with a specific biotechnology briefing – the GBF in Braunschweig-Stockheim (see West Germany section below). In other countries, information on biotechnology R & D must be gained from a scattered array of sources. Although there are biochemistry societies and institutes in all countries, the first port of call is often the trade associations representing the chemical or pharmaceutical industries. These are often very influential associations due to the strength of the chemical and drug companies, especially in West Germany, the UK and Switzerland.

6.4   National Profiles

This section gives reviews of public and private developments in biotechnology, with further sources of information, in the major European countries.

WEST GERMANY

West Germany has taken the lead in European biotechnology for two important reasons – it is the largest and most stable industrial economy in the region, and it is the host country to the region's largest chemical and pharmaceutical enterprises. These companies, notably Bayer and Hoechst, have dominated biotechnology research and investment, and this has restricted the development of small, venture-capital companies, which have characterised development in the USA and the UK. However, Germany has also had the strongest public policy toward biotechnology, from both the federal government and individual states.

Public Policy

The federal government has played a leading role in biotechnology since 1980, through the BMFT (Federal Research and Technology Ministry). The BMFT supports government policy which is aimed at ensuring an even balance between public and private investment. At present, annual investment in pure research amounts to about $500 million a year, of which government finds 30 per cent and private industry 70 per cent.

A major priority of the government is to encourage partnerships between universities and the industry. In line with the 'heavy' industrial interests of Germany's large chemical companies, the BMFT has shown particular interest in industrial and agricultural applications in the drug and food industries. Funded projects of this kind include:

waste disposal organisms
enzyme bioreactors
biological pest control
raw materials from waste
production of biogas and bioalcohol

For the long term, the BMFT will also fund research into the risks of biotechnology for society, and will encourage German scientists not to emigrate to the USA. Cooperative agreements with institutes in other countries are also considered.

With a 1984 budget of DM 60 million, drawn from government funds, the Biotechnology Research Institute (GBF) is a major contributor to German biotechnology, and the only such specific institute in Europe. The 120 scientific staff are spread among ten divisions, and this indicates the breadth of R & D undertaken by the GBF, covering all major areas of biotechnology. Specific areas of ongoing work include:

restriction endonucleases
saccharification of cellulose materials
production of steroids with microbes

Contacts:

Federal Research and Technology Ministry:
Bundesministerium für Forschung und Technologie (BMFT)
Biology and Non-nuclear Industry Division
Postfach 200706, D-5300 Bonn 2

Biotechnology Research Institute:
Gesellschaft für Biotechnologische Forschung (GBF)
Mascheroden Weg 1, D-3300 Braunschweig-Stockheim

Other related organisations are the European Molecular Biology
Organisation (Heidelberg), and the Federation of Chemical
Industries (Frankfurt).

Private Companies

The German biotechnology industry has been characterised by an
absence of small research-oriented biotechnology firms. Instead,
research is conducted in-house by the chemical giants led by
Bayer and Hoechst, but all of which feature in Table 3 (above) of
Europe's largest chemical concerns:

Bayer AG (Leverkusen-Bayerwerk) was a pioneer in biochemical research a century ago, and the company is particularly strong
in forging links with research institutes. One of the most
significant of these is a cooperative effort with the Massachusetts
Institute of Technology aimed at molecular biology. Another is a
link with the German Max Planck Institute. Enzymes, blood
products, MABs and diagnostics are also named by the company
as targets. Commercialisation of discoveries on an international
scale should eventually take Bayer to an eminent position as a
supplier of many biotechnological products. The company
already has large subsidiaries throughout the world; most notable
of these are the Miles and Cutter Laboratories in the USA.

Hoechst AG (Frankfurt) is, like Bayer, one of the world's five
largest pharmaceutical companies, and yet drugs form only a
small part of the group's total business. Biotechnology was given
priority at Hoechst as early as 1971, with a strong focus on genetic

engineering and the production of interferon. Hoechst's largest outside arrangement has been with the Massachusetts General Hospital (under Harvard), comprising a $50 million grant spread over ten years to finance basic research; resulting products will, of course, be licensed to Hoechst. Hoechst also has subsidiaries in the USA and Japan; its most important European subsidiary (from the biotechnology angle) is in Roussel-Uclaf in France.

Boehringer Ingelheim (Ingelheim am Rhein) is another major pharmaceutical company, strongest in Germany itself and in Italy, which has concentrated research in the immediately commercial area of interferon. Basic research in the bio-chemical area has been conducted in Ingelheim since 1962. Boehringer Mannheim, a completely separate company, has specialised in MABs. Both of these companies have subsidiaries in the USA and elsewhere in Europe.

Schering AG is yet another major German chemical/pharmaceutical firm which is forging ahead with long-term biotechnological research. A ten-year fund has been set up with the City of Berlin to finance genetic engineering research (jointly worth $33 million), and in the USA Schering is working with the Boston Genetic Technology Institute.

Degussa AG (Frankfurt) is a specialist in amino-acid production and has ongoing enzyme research of a general nature. Also involved in German biotechnology are E. Merck, Gruenenthal, and BASF, all very large chemical concerns. The most notable of the few specialist companies is the interferon specialist, Bioferon Biochemische Substanzen GmbH in Laupheim.

FRANCE

Like many other European countries, France made a slow start in investing in biotechnology but is attempting to catch up with other developed countries. Political volatility has not helped in producing coherent, long-term government policies over the last five years. In particular, the country's largest chemical and pharmaceutical concern, Rhône-Poulenc, has been nationalised and re-privatised within the last decade, and the medical establishment in France has been convulsed by political infighting.

Public Policy

A very precise commercial target has been set for France's

industry – a 10 per cent share of all commercial sales of biotechnology products in the world by the early 1990s. This is just one aim of a 'mobilisation programme' intended to make France the leading European science and technology nation by the end of the century. Among the very practical goals of the policy are the intention to exploit biotechnological sources as a means of reducing French dependence on imported oil. This policy will obviously influence, or be implemented by, the partly-public Société Nationale Elf-Aquitaine (SNEA).

The divisions of the government budget, to be spread among various organisations, are an indication of French priorities. Three research areas have received annual grants of FF7 million – classic biology, microbe production, and logistical support; five areas are worth FF4–5 million each – genetic engineering, vaccines, bioreactants, fermentation technology and enzymes; while FF3 million has been allocated to plant improvement and FF2 million to education in biotechnology. In more general terms, a logical decision has been taken, given the structure of the French economy, to emphasise research in the food and agricultural areas and in immunology.

The Ministry for Research and Industry is the ultimate source of most public funds for both private industrial grants and grants to universities and research institutes. However, in France as in most other countries, there are several reputable, rival agencies either crying out for funds or attempting on their own to co-ordinate research policy. These include the Guidance Committee for Strategic Industries (CODIS), which has encouraged nearly $1000 million to be invested in French private biotechnology companies since 1981. Other organisations involved, with policies evident from their names, are the National Agency for Utilisation of Research Results (ANVAR) and the National Institute for Agronomy Research. Academic research is coordinated by the National Centre for Scientific Research (CNRS). The specific organisations in biotechnology are the Federation of Biochemical Industries, and the Association for the Development of Biotechnology. France also has several famous research institutes, especially the Institut Pasteur and the Institut Merieux, discussed below.

Contacts:

Ministry for Research and Industry

Ministère de la Recherche et de l'Industrie
'Biotechnology Mission'
5 rue Descartes
76005 Paris

Association for the Development of Biotechnology (ADEBIO)
Association pour le Développement de la Bio-Industrie
3 rue Massenet
77300 Fontainebleu

Federation of Biochemical Industries
Syndicat de l'Industrie Chimique Organique de Synthèse
et de la Biochimie
57 avenue Marceau
75116 Paris

Private Companies

The French chemical industry is dominated by the country's major multinational, Rhône-Poulenc, a company of international standing. This outstanding position helped make Rhône-Poulenc an obvious target for nationalisation in the late 1970s. Despite changes in government attitude to the status of the company, it has been possible to develop a reasonably cohesive forward strategy in biotechnology. Via a more complicated ownership route (see below), the state oil company, Elf-Aquitaine, is also deeply involved in biotechnology.

Rhône-Poulenc SA (Courbevoie) has traditionally been strong in pharmaceuticals (notably vitamin B12, antibiotics and anticancer agents) and in chemicals for the massive French agriculture industry. However, as the country's 'national' chemical concern (state owned or otherwise), Rhône-Poulenc is involved in most major sectors of the chemical industry. Rhône-Poulenc also operates or invests in smaller companies which have a more specific research brief. These include Laboratoire Roger Bellon (vaccines, cancer agents), Lacto-Labo (fermentation products for the food industry), and the Institut Merieux. The subsidiary Genetica is devoted entirely to genetic engineering, particularly in agricultural areas (plant tissue culture, hybrids, food additives, fertilisers). With its 65 per cent shareholding from Rhône-Poulenc, Genetica is probably one of Europe's most stable specialist biotechnology companies.

The Institut Merieux, like Pasteur, has a typically French structure in involving many private investors (including the Merieux family itself) together with a 40 per cent shareholding held by Rhône-Poulenc. Merieux is a major commercial producer of vaccines and blood derivatives as well as being a research institute. Interferon is being produced in Lyon, and the Institut has a joint research agreement with Genentech (USA), the world's largest specialist bio-company.

Elf-Aquitaine (Paris, La Défense), the state-controlled oil company, is directly involved in biotechnology research through investments which include Elf-Bioindustrie, Sanofi, and Institut Pasteur. The parent company is naturally most interested in the subject of biomass conversion for alternative energy sourcing, plus oil recovery and other industrial applications of biotechnology. An advantage of Elf is that it has been able to work easily with another state-controlled company, the chemical company Entreprise Minière et Chimique. EMC's relevant interests are in agricultural chemicals, with bacterial nitrogen fixation its major project.

Sanofi (Paris) is Elf's pharmaceutical arm. The company is well advanced in biotechnology, already manufacturing interferons, researching hormones, vaccines, fermentation and immunology applications, and working with jointly-held subsidiaries in the USA (American Home Products) and Japan. The company has the advantage of owning 51 per cent of the Institut Pasteur Production. IPP is the commercial arm of the Institute (which owns the other 49 per cent), specialising in R & D in vaccines, sera, blood products, diagnostics and interferon.

Several other French establishments are worthy of note. Roussel-Uclaf is a major drug company, partly owned by Hoechst of Germany, which claims to be working in a broad spectrum of research areas of biotechnology. Lafarge Coppée is a cement manufacturer which incorporates a Biochemical Group of small research companies – notably Orsan and Eurolysine – specialising in agricultural research. At the other end of the scale are the prominent specialists, Transgene SA and Immunotech SA. Transgene (Strasbourg) was set up like Celltech of the UK in 1980, utilising funds from many sources: Banque Paribas, Assurances Générales, the food/brewing giant BSN, Elf, Sanofi, Institut Pasteur, and drinks giant Moet-Hennessy. Small shareholdings were also distributed to universities and research institutes. Rapid gene-synthesis is an area of major interest, with agricultural end-uses a high priority. Immunotech, in Marseilles, is a much smaller organisation, but with more precise research targets in hybrid technology, and once again with strong financial backers (Elf and Compagnie Française des Pétroles).

UNITED KINGDOM

A long-term view shows that the UK has declined as an industrial power, certainly in comparison to the progress made by Japan, West Germany and France. Public funding has been cut back in many sectors of the economy, with the 1979–1986 Conservative government putting the onus on private investment as the way forward. These 'cuts' have been particularly severe in the academic sector, and this has contributed to a further increase in the 'brain drain' whereby eminent scientists emigrate to the USA in search not only of higher personal remuneration but also of adequate research facilities and funds. On the positive side, the UK boasts four strong multinational drug companies – Glaxo, Wellcome Foundation, Beecham Group and Imperial Chemical Industries (ICI). These companies have mainly been on the ascendant in the 1980s, with exports and foreign business increasing in value, and they have been joined by The Boots Company, which has played a primary role in biotechnology by working with Celltech, the UK's major specialist company.

Public Policy

The UK has yet to set up a prominent central administrative unit for biotechnology. As in France, various organisations compete for government grants. These have an academic bias, being channelled through the Department of Education and Science (University Grants Committee, Science Research Council). The Department of Trade and Industry made its first substantial foray into biotechnology in 1980, by helping to fund the setting up of the commercial company, Celltech, with an $11 million investment through the British Technology Group (see Celltech profile below). The Group also founded the Agricultural Genetics Company Ltd, a private company, in partnership with Ultramar Oil. AGC has rights to commercialise projects emanating from research undertaken under grants from the Agricultural Research Council, in the same way that Celltech has first refusal on discoveries made under the auspices of the Medical Research

Council. A further $10 million a year of central government funds has been made available to academic establishments and research institutes, of which the most prominent is the Centre for Applied Microbiology and Research (CAMR) at Porton Down.

Attempts have been made to create organisations with a specific interest in biotechnology. These include the British Coordinating Committee for Biotechnology and the Association for the Advancement of British Biotechnology (AABB). Public and private research priorities have been clearly in the medical field, led by Celltech's commercialisation of monoclonal antibodies and, through Boots-Celltech, of diagnostic kits. Agriculture is another focus of research.

Contacts:

British Technology Group
101 Newington Causeway
London SE1 6BU

Association for the Advancement of British Technology
c/o Food Research Institute
Colney Lane, Norwich NR4 7VA

British Coordinating Committee for Biotechnology
CAMR, Porton Down
Salisbury, Wiltshire SP4 0JG

Private Companies

Celltech and its partner, Boots-Celltech, have led the way in pioneering commercialisation of British biotechnology. As indicated in the profile earlier in this chapter, Celltech is now one of Europe's pre-eminent specialist biotechnology companies, and is set to break even in 1987–1988. Apart from Celltech, the emphasis in the UK has been on the creation of numerous venture-capital companies – more than anywhere else in Europe – although financiers in the City of London were more enthusiastic in the late 1970s than they have become since. Profitability at Celltech could well encourage a resumption of this investment interest. The Agricultural Genetics Company has risen to prominence in 1987 with the announcement that it has patented a gene enabling some plants to become pest-resistant.

Porton International is another specialist company which could become an international leading specialist in commercial biotechnology. Started in 1983, Porton already has capital of over $500 million, and comprises ten companies involved in various aspects of the industry and on both sides of the Atlantic. The company has the right to commercialise products developed at the Porton Down centre (CAMR, see above), following a familiar pattern in British biotechnology development (i.e. Celltech with the Medical Research Council, and AGC with the Agricultural Research Council). Other prominent specialist research companies, although mainly without the commercial potential of the three mentioned above, are IQ (Bio) at Cambridge, the Leicester Biocentre, CLEAR (Cambridge Labs. for Energy and Resources), Invaresk and Sera-Lab. The UK is also not short of investment (venture-capital) companies, such as New Market and Biotechnology Investments Ltd. The latter has money tied up in Celltech and in other European biotechnology specialists.

The UK's major chemical and pharmaceutical companies all have their own biotechnology research plans, as a natural outcome of their traditional strength in antibiotics (Beecham Group and Glaxo) and antivirals (Wellcome Foundation). Having been for most of its history a private, research-oriented foundation, Wellcome is seen as having among the best chances of producing biotechnological products on a commercial scale, and has been attracting investment and attention for its research into the AIDS virus. However, Glaxo is also attracting considerable international investment, and is one of the fastest rising pharmaceutical multinationals of the 1980s. Beecham Group is restructuring, but still has strong biotechnology-relevant divisions in several areas — ethical and household medicines, toiletries and cosmetics, household chemical products and food and drink.

Imperial Chemical Industries (ICI) is also a major drug producer, but is best known as Britain's largest general chemical company, dominating that industry almost as much as Rhône-Poulenc does in France. The giant Anglo-Dutch oil company Shell is also investing in biotechnology, as are several other major UK consumer-product companies including Unilever, Tate & Lyle (sugar) and Ranks Hovis McDougall (bakers, foods). To add to these, the UK boasts most of Europe's largest brewing and distilling groups because the alcohol industry is very concentrated, and these companies (Guinness-Distillers, Bass, Imperial, Allied-Lyons) have both the resources and the inclination to invest in related research.

OTHER COUNTRIES IN EUROPE

Switzerland

Biogen, as already profiled, was the first of the major European specialist companies to be founded, and the company has continued to be a leader in various fields in the 1980s. Other specialist companies are Uniserum, PetroGenetic and Chemap. For a small (but prosperous) country, Switzerland has an outstanding selection of very large chemical/pharmaceutical companies on a par with those in Germany and the UK. These will dominate Swiss involvement in biotechnology in future, working closely with publicly funded organisations and continuing to spend a high proportion of their income on R & D.

During the 1980s, Ciba-Geigy has developed a specific biotechnology R & D division covering all major areas of research in line with the parent company's diverse chemical interests. Sandoz, a more specialised drug multinational, has concentrated on genetic research, and has strong American links – a share in Genetics Institute (Boston), its own Sandoz Inc. subsidiary, and an agreement with the Wistar Institute (Philadelphia) on MAB research.

Hoffmann-La Roche is also specialising in genetics, and has long-standing agreements with the Basle Institute for Immunology, and with Genentech (California) and with the top Japanese drug company Takeda on interferon research and production.

Public policy is complex and fragmented, as there is traditional rivalry and competitiveness between the linguistic regions (French and German) and among the cantons (member states of the Swiss Federation), as well as between the large chemical companies, and the universities. Swiss academic prowess is well known, producing high standards of research and qualified staff in the universities of Geneva, Lausanne, Zurich, Bern and Basle. Federal unity is provided by the Federal Institute of Technology in Zurich (see list below of 'Public Organisations – rest of Europe'). However, the sizes of budgets and funds are closely guarded secrets in both public and private sectors.

Scandinavia

Denmark and Sweden dominate Scandinavian biotechnology because of their size and industrial strength ahead of Norway and

Finland. Both Denmark and Sweden are host to companies with strong specialist positions in international pharmaceuticals. In Denmark, Novo Industri and Nordisk Gentofte are Europe's two largest suppliers of insulin, specialising in modified porcine insulin. The Swedish state-owned Kabi Vitrum is reknowned worldwide for human growth hormone (HGH).

Novo Industri has marketed a genetically engineered human insulin in competition with Eli Lilly in the USA and in Europe; the company has R & D links with Biogen (Switzerland) and Takenaka (Japan), and a production agreement with Squibb (USA). Novo also has its own subsidiaries in North America and Japan. Research has spread from basic genetically-altered insulin to other rDNA applications in enzymes, peptides, interferon and hybrids. Nordisk Gentofte is a major producer of insulin and HGH. The company has worked with Chiron in the USA and with Wellcome in the UK on porcine insulin projects. Denmark's major brewery, United Breweries, has been one of Europe's first such companies to set up a special division – Carlsberg Biotechnology, working in protein manipulation and peptide synthesis. Other work is being carried out by Danish Sugar Refineries (plants and agricultures), Hansen Laboratories (food), and Dakopatts (MABs, diagnostics). Hansen and Dakopatts have USA subsidiaries and R & D agreements with other specialist companies outside Denmark.

Kabi Vitrum set up a subsidiary, Kabi Gen, to carry out research in interferon and blood applications of biotechnology, while the parent company has agreements to market Lilly's insulin in Scandinavia and has R & D agreements with Genentech. The other major Swedish company is Ab Fortia, which is developing new bacteria and reagents, and has acquired an American subsidiary as well as working with Genex. It is expected that Sweden's largest private drug multinational, Astra, will also play a significant role in future biotechnological development in Sweden.

Government backing works closely with industry in Scandinavia because of the traditional public ownership of large, strategic companies like Kabi Vitrum, Alko Oy (the Finnish alcohol monopoly) and the energy companies. Central funding in Sweden is the responsibility of the National Board for Technical Development, which has financed a $20 million bio-lab at the University of Stockholm and given annual grants to organisations totalling over $25 million since 1981. In the rest of Scandinavia, public funding is on a much smaller scale. The main problem faced is a shortage of qualified biotechnologists.

Benelux Countries

Belgium-Luxembourg and the Netherlands have experienced different problems in developing biotechnology. In Belgium, the decline of heavy industry has tightened budgets all round, while in the Netherlands problems were caused by a delay in liberalisation of laws on genetic engineering. This was strongly influenced by environmentalist lobbies, which have considerable influence in the crowded Benelux region when it comes to setting up technical laboratories and factories.

Akzo is the largest chemical/pharmaceutical company of the Benelux region. (Belgium hosts two major chemical companies – Solvay and UCB – but neither of these ranks as a major drug producer.) In 1982, Akzo's subsidiary Intervet began to market an rDNA vaccine for veterinary use, several years before the marketing of most other biotechnology products was even considered. Akzo has continued to work with Collaborative Genetics (USA) on rDNA applications in insulin and vaccine production. Gist-Brocades is another company with strong potential in Dutch biotechnology, as it produces a quarter of the world's crude penicillin, and is now concentrating on large-scale enzyme fermentation developments. Large-scale industrial production potentiality is a feature of Dutch biological research, based around the very large industrial companies operating from the Netherlands: Royal Dutch Shell, Unilever (Anglo-Dutch), Petrofin, and Heineken, Europe's largest brewer.

In Belgium, economic constraints have limited investment in biotechnology, and it is probably too late for new companies to develop significantly in the medium term. Unlike the Netherlands, however, there are numerous small specialist companies involved in specific bio-applications. Solvay, the largest chemical company, leads the way, specialising in agricultural applications, and the second largest company owns UCB-Bioproducts. As in Switzerland, public funding is fragmented in Belgium, with two rival sources coming from the Wallony Regional Executive (French-speaking) and the Flemish Industrial Development Agency (Dutch-speaking). These regional bodies have set up their own research subsidiaries. In the Netherlands, a modest government investment plan is in place, but it has also been necessary to set up a committee to examine ethical and moral issues, and this climate may deter foreign companies from choosing the Netherlands as a host for investment.

Italy and Spain

Public funding in Italy is always hard to generate, and only small amounts have been allocated to specific biotechnology projects. Private industry is hamstrung by the lack of strong patent laws. This situation has made for a weak international position in pharmaceuticals, where biotechnological products lie. By far the largest Italian company in a field dominated by foreign companies is Montedison, the chemical combine which owns Farmitalia Carlo Erba, Italy's largest drug company. Specialist companies include Biotec of Bologna, Serono, and Sorin Biomedica. Public research institutes such as the Institute of Biochemical and Evolutionary Genetics rely on grants from the National Research Council.

Funding in Spain is even lower for biotechnology, coming from the Industrial and Technical Research Council. Spain lacks any large multinational chemical companies with major research resources. The main specialist company is Immunologia y Genetica Aplicado SA, set up and 51 per cent owned by the Spanish government.

6.5 Outlook

In the constant debate over the future of the biotechnology industry, a consistent theme has been the balance in international power held by the three main developed economic regions – the USA, Japan and western Europe. The USA and Japan have moved faster than most European countries, but this is to ignore the fact that industrial development in the individual countries of Europe has taken place at very different speeds. The concept of Europe as a homogenous economic region is a false one, although it is convenient for American and Japanese companies (and their governments) to use as a yardstick of their own national development.

Western Europe acts as the headquarters for at least a dozen large and innovative multinational companies. The activities of these companies, such as Bayer, Hoechst, ICI, Rhône-Poulenc,

TABLE 6.11   CONTACTS: PUBLIC ORGANISATIONS IN BIOTECHNOLOGY IN THE REST OF EUROPE

Switzerland	Federal Institute of Technology (Zurich) (Eidgenoessische Technische Hochschule)
	Federal Office for Industry and Trade (Bern)
	Swiss Union of the Societies for Experimental Biology – UGEB (Basle)
	Swiss Association of Chemical Industries (Zurich)
Scandinavia	
– Sweden	National Board for Technical Development (Stockholm)
	Foundation for Biotechnical Research (Stockholm)
	National Science Research Council (Stockholm)
	Association of Swedish Chemical Industries (Stockholm)
– Denmark	Danish Council for Scientific Policy and Planning (Copenhagen)
	Department for Development – DANIDA (Copenhagen)
	Institute of Biological Chemistry (Copenhagen)
– Finland	Finnish Council of Scientific and Technological Information (Otaniemi)
Benelux	
– Netherlands	National Central Organisation for Applied Science Research – TNO (The Hague)
	Netherlands Biotechnology Society (Delft)
– Belgium	Ministry of Economic Planning and Scientific Policy (Brussels)
Italy	National Research Council (Rome)
	Directorate of Research and Development (Rome)
Spain	National Council for Research and Development (Madrid)
	Industrial and Technical Research Council (Madrid)

Source: Authors research

and Ciba-Geigy, transcend national boundaries. Even more than their American and Japanese counterparts, these chemical companies are truly multinational in scope. As is now the case in the pharmaceutical industry, it will be necessary to launch technological products to a world market in order to recoup the cost of initial investments (in plants, staff, patents). This is of particular urgency to large companies whose domestic markets are limited in size – e.g. Akzo of the Netherlands, Novo in Denmark, and all the Swiss companies.

In the secondary stage of the biotechnology industry, which can be dated from 1983–1990, these companies have been busy either creating biotechnology subsidiaries in the USA and Japan, investing in existing research-based companies in those countries, or forging mutual agreements on R & D and commercialisation with American and Japanese companies and research institutes. This means that the national frontiers are becoming increasingly confused. By the year 2000, biochemical products could be commonly available to industry and the consumer which have been 'invented' in one country, developed by a company in another country, and commercialised by yet another company in a third country!

MARKET PROSPECTS

Given that the shifting economies of twenty European countries must be considered, it is virtually impossible to forecast the future size of the markets for bio-products in this region. However, the current situation is known to the extent that the western European market accounts for between 25 per cent and 30 per cent of virtually every industrial product and process by their value of sales. This ratio can be expected to continue into the next century.

The world market for all products of biotechnology has been estimated and forecast by various organisations. A US government compilation of these surveys showed that forecasts for the world market in 2000 range from around $50,000 million to $150,000 million. The range is less extreme for health care products – from $10,000 million to $23,000 million – but it is anyone's guess of the commercialisation of agriculture and food products, and the potential for biological energy products is an even greater unknown. The following table therefore estimates the possible range of values of European markets, in order to indicate at least that in the medium term it is food and agriculture enhancements which appear to have most potential.

An important factor to bear in mind is that European regulations will, by the year 2000, be much more influenced by European Community policy than they are now. However, member state governments (and those of non-members) will still pursue various policies with regard to genetic engineering and biotechnology. It will be surprising if conflict does not develop within the Community as the full potential for change of biotechnology makes itself known to a wider (voting) public. In

TABLE 6.12   POTENTIAL MARKETS FOR BIOTECHNOLOGY APPLICATIONS IN WESTERN EUROPE IN THE YEAR 2000

	$ million
Health care, drugs	5000 – 10000
Agriculture, food	5000 – 30000
Energy	2000 – 5000
Industrial chemicals	2000 – 3000
Total (low and high forecasts)	14000 – 48000

Source: Authors forecasts

their acceptance and rejection of new 'material' innovations, Europeans can, in general terms, be said to be far more conservative than Americans or Japanese, although they are more open to less tangible cultural changes.

Biotechnological innovation in health care is perhaps the least likely to encounter opposition since the first products are unlikely on the face of it to be dangerous. Indeed, purer insulin and antibiotics and more effective diagnostic kits clearly offer advantages to a European public already accustomed to innovations in medical treatment once unheard of (abortions, organ transplants and so forth).

The reaction to genetic engineering for the purpose of producing high-yield crops and animals and for manipulating the environment is likely to be very different in some countries (especially Germany, where the Green (ecology) Party is firmly established as a political force). In food and agriculture, the European trend is toward more 'natural' whole-foods, and the use of EEC determined guidelines on food additives is spreading. This will at least mean that suspicion will greet many new biotechnological products.

At present, the low price of crude oil has taken the impetus out of the alternative-energy industry, but biotechnology still offers a neat long-term solution to the problem of finding sources of energy which are cheap, and which replace themselves, as well as getting rid of waste matter. In the long run, the availability of much cheaper and more efficient alternatives to oil and coal will totally disrupt the way the world economy is shaped. This will have particularly noticeable effects in Europe, where there is currently a clear gulf between the oil-rich and oil-less countries.

INDUSTRY PROSPECTS

There seems little doubt that those countries and companies which have invested substantially in biotechnology in the early 1980s will reap ample rewards in the long term, despite the short-term disappointments. This means that the gap between the richer economies – mainly in north and central Europe – and the poorer countries – southern and western – could increase substantially in the twenty-first century. However, biotechnological products will probably be made available in sufficient quantities for all countries, by mainly multinational companies, and this will reduce the economic gap.

With a strong, stable and growing economy, West Germany is likely to remain the European leader, and to increase its lead both scientifically and commercially. When the products come out of the laboratories, only multinationals with an existing network of production and marketing subsidiaries around the world will be able to take full advantage. This points squarely at the Bayers, Hoechsts, Ciba-Geigys and ICIs of Europe's chemical industries. The multinational oil companies, ever on the watch against being usurped by new energy forms, will undoubtedly be at the forefront as well.

Smaller research-based companies should, however, continue to be in demand. There is no indication that the well has run dry as far as biological discoveries are concerned. Indeed, a historical perspective from the viewpoint of the next century will probably show that the biotechnology industry was barely out of its infancy even in the late 1980s. A continuous process of discovery and development in biotechnology can be expected for the foreseeable future, and this will support the existence of the new breed of 'laboratory' enterprises like Biogen, Celltech and Genetica, although a dozen or so European multinationals will end up reaping the most benefits from the biotechnology revolution in the purely commercial terms of mega-dollar product sales.

Part Three
Issues and Impacts

7 Chapter Seven

Regulatory issues in biotechnology

7.1 Introduction

The introduction of genetic engineering techniques to biotechnology raised new issues which promoted much public discussion. Early researchers raised a number of potential concerns at the Asilomar Conference in the USA in 1975 which led to the establishment of control mechanisms for research involving recombinant DNA in a number of countries. The USA set up the Recombinant DNA Advisory Committee (RAC) whose early work involved defining review procedures (including the setting up of local Institutional Biosafety Committees), establishing guidelines for risk assessment of the potential hazards believed to be associated with genetic engineering, and providing guidance on appropriate levels of physical containment within laboratories for different types of organisms and experiments. Similar bodies were set up in a number of countries (e.g. the UK set up the Genetic Manipulation Advisory Group – GMAG), and these generally provided guidance on similar lines to the RAC. In most cases, such guidelines were, as with the RAC, voluntary, but were supported by the scientific community as providing a framework in which research could proceed under the supervision necessary to meet the concerns of the public, employers, employees and government departments.

As experience was gained and the potential risks could be quantified better, many of the initial guidelines were relaxed by RAC and GMAG and by other national bodies, and many scientists now regard the specific safeguards on rDNA work as part of the much wider framework of good safety practices for biological laboratories. By the early 1980s, however, the RAC and analogous bodies were being faced with very different kinds of regulatory considerations. Genetically manipulated organisms were being used in large-scale industrial processes whose products were reaching the public, and modified organisms themselves were under consideration for use as plant growth promoters or pesticides involving the planned release and growth of genetically modified organisms in the environment. There has thus been an increasing amount of attention given in recent years to these aspects arising from the industrial applications of biotechnology.

Most countries have started from a position of having regulations over the products of traditional biotechnology and have recognised that the 'new' biotechnology essentially applies different (and much more precise) techniques for optimising the properties of micro-organisms for industrial purposes. There is thus a predisposition to regulate the product irrespective of the production process, which in general favours the use of existing regulations rather than creating new ones. Of particular importance are the standards which have evolved in the pharmaceutical industry to deal with the large-scale culture of micro-organisms under the general heading of Good Manufacturing Practice (GMP). These set general requirements for manufacturing establishments and cover subjects such as equipment standards, personnel procedures, record-keeping, quality control, operation and maintenance. These practices are also generally applicable to the culture of recombinant organisms.

National Approaches to Regulations

USA

The USA has played a leading role in the evolution of the guidelines covering rDNA research since the RAC issued its first guidelines in June 1976. These subsequently formed a model for national guidelines in a number of other countries. The guidelines were enforceable only for researchers receiving National Institute of Health funding, but have been widely adhered to both in universities and in industry voluntarily. Early restrictive conditions were progressively relaxed in the light of the latest scientific knowledge, and many of the simpler genetic manipulations involving a range of host-vector systems are now cleared at the local level by the Institutional Biosafety Committee, with the RAC's attention reserved for the more problematical proposals or matters involving new applications of the technology (e.g. with deliberate release) or general policy.

Over the years the RAC evolved into the central body of expertise on recombinant DNA regardless of the product or area of application under consideration. While its composition was suitable for advising on research, it was less appropriate for it to consider, for example, agricultural applications. This issue reached a turning point when a federal court ruled in 1984 that the NIH had violated the National Environmental Policy Act by failing to prepare an Environmental Impact Statement for its 1978 guidelines on deliberate environmental release, and for the specific permission granted to the University of California for the field testing of a genetically engineered bacterium.

This was followed by a proposal by the President's Office of Science and Technology Policy for a Coordinated Framework for the Regulation of Biotechnology which would clarify the role of the various agencies involved. This was finally published in June 1986 and laid out the responsibilities of the Food and Drug Administration (FDA), the Environmental Protection Agency (EPA), the US Department of Agriculture (USDA), the National Institute of Health (NIH) and the Occupational Health and Safety Administration (OHSA). Mention was also made of the potential of the Export Administration Act to cover biotechnology. The publication in the Federal Register provided a 'matrix' which defined the responsibilities of each agency and identified areas of overlap. This is summarised in Table 7.1.

To ensure that there would be a consistency of approach between the various agencies in regulating biotechnology, a Biotechnology Science Coordinating Committee (BSCC) was set up comprising the National Science Foundation (NSF), NIH, FDA, USDA and EPA. It will identify gaps in scientific knowledge, promote consistency in the development of agencies' review procedures and assessments, and serve as a coordinating forum for addressing scientific problems, sharing information and developing consensus. The BSCC does not replace the RAC but now allows that body to restrict its considerations to the biomedical area. Other agencies have made their own arrangements for in-house advice and expertise on biotechnology.

Products of biotechnology are thus regulated according to existing legislation administered by the appropriate agency

**TABLE 7.1   US COORDINATED FRAMEWORK FOR THE
REGULATION OF BIOTECHNOLOGY – AGENCY
RESPONSIBILITIES**

Subject	Responsible Agency
Foods/Food additives	FDA[1], FSIS[1]
Human drugs, medical devices, biologics	FDA
Animal drugs	FDA
Animal biologics	APHIS
Other contained uses	EPA
Plants and animals	APHIS[1], FSIS[1], FDA
Pesticide micro-organisms released into the environment	EPA[1], APHIS[3]
Other uses	
Intrageneric combination	EPA[1], APHIS[3]
Intergeneric combination	
Pathogenic source organism	
1. Agricultural use	APHIS
2. Non-agricultural use	EPA[1], APHIS
No pathogenic source	EPA report
Non-engineered pathogens	
1. Agricultural use	APHIS
2. Non-agricultural use	EPA[1], APHIS
Non-engineered nonpathogens	EPA report

Notes:
1. FSIS, Food Safety and Inspection Service is responsible for food use.
2. FDA is involved when related to a food use.
3. APHIS is involved when the micro-organism is a plant pest, animal pathogen or regulated article requiring a permit.
Source: Lead Agency

identified in Table 7.1. The FDA regulates all drugs, diagnostic kits and biologicals for human use, and evaluates the products of rDNA using the same review procedures as for other products. The EPA has responsibility for pesticides approval under the Federal Insecticides Fungicide and Rodenticide Act (FIFRA) and has already dealt with the applications for genetically engineered microbial pesticides. While these are generally handled via the same procedures as before, some specific changes to the regulations have been made for genetically engineered products – in particular in withdrawing the 10-acre exemption for small-scale field trials. An experimental Use Permit is now needed from EPA before any field trial of a genetically engineered pesticide can take place.

144

The Environmental Protection Agency has a potentially broader statutory authority to control biotechnology through the Toxic Substances Control Act (TSCA) which requires the pre-manufacture notification (PMN) of new chemicals. In the Federal Register notice on EPA's policy, it was stated that TSCA would apply to all genetically engineered organisms used for commercial purposes. The commercial use of a new organism would thus be subject to the usual PMN system. EPA stated, however, that they would focus their regulatory resources on three categories of organisms:

Micro-organisms that contain new combinations of traits (particularly intergenetic organisms)

Micro-organisms that are pathogenic or contain genetic material from pathogens

All micro-organisms that are subject to deliberate release into the environment

The US Department of Agriculture regulates plants and plant products, plant pests, noxious weeds, and viruses, sera, toxins and other products intended for the treatment of animals. It has a number of departments that handle biotechnology, including the Animal and Plant Health Inspection Service (APHIS), which is responsible for protecting the nation's animal and plant resources from disease and pests; and the Agricultural Research Service which together with the Cooperative State Research Service is responsible for USDA's research efforts.

Although some overlap in responsibilities still exists (e.g. between FDA and USDA on animal health, and between USDA and EPA on pesticidal micro-organisms released into the environment), lead agencies have been identified with the objective of avoiding regulatory delays caused by ambiguity in the statutes and the agencies' policies.

UNITED KINGDOM

In the UK, the regulatory authorities have already approved recombinant insulin, human growth hormone and alpha-interferon under the Medicines Act (1968) – the legislation used to control all human drugs. A novel food based on a non-genetically engineered mycoprotein has been marketed under authority of the Food Act, and biorational pesticides approved for use under the pesticides legislation (The Food and Environment Protection Act 1986). Most of the products currently under development are expected to be regulated under existing legislation without the need for any major changes, but it has been necessary to develop mechanisms to provide more specific guidelines on the data needs and procedures for regulating the products of genetic engineering.

The UK's mechanism for providing this specialist advice is via the Advisory Committee on Genetic Manipulation (ACGM), which replaced the GMAG in 1984. The ACGM's primary responsibility is to advise the Health and Safety Commission and Executive under the Health and Safety at Work Act 1974 – the UK's primary legislation on the protection of workplace health. ACGM also advises other departments on matters concerned with genetic manipulation, thus avoiding the unnecessary duplication of specialist committees throughout departments and ensuring that available technical expertise is applied efficiently and consistently. The committee has an independent Chairman, five representatives of employers, five of employees, and eight which are selected on the basis of their scientific or medical speciality. The only legislation to have been specifically introduced to deal with genetic engineering has been the Health and Safety (Genetic Manipulation) Regulations 1978, which require the notification to HSE of any intention to carry out genetic manipulation as defined and the provision of the details of experiments.

On its formation in 1984, the ACGM initiated a review of the issue of deliberate release and issued guidelines on the planned release of genetically manipulated organisms into the environment in April 1986. These required the prior notification of all proposals to the ACGM before any work took place. The scheme as introduced was voluntary, but consideration is being given to amending the relevant section of the Health and Safety (Genetic Manipulation) Regulations of 1978 to make such prior notification mandatory. The guidelines recognise that it is not possible at the present to specify an inflexible list of data requirements that must be included in the risk assessment. However, they have identified a number of factors that are likely to assist the proposer to make his initial risk assessment, and these should normally be addressed in applications where relevant.

The ACGM has also nearly finalised guidelines on the large-scale industrial use of genetically engineered micro-organisms, and these are expected to be published shortly, and to be consistent with the criteria in the OECD report (see later). The UK is also relying on existing legislation to regulate biotechnological products in the food and agricultural areas. Foods and food additives are controlled by the Ministry of Agriculture, Fisheries and Food under the Food Act (1984), pesticides must be approved under the Food and Environment Protection Act (1986), and the Ministry also has responsibilities for the control of plant pests. Primary regulatory authority remains with the Ministries when genetic engineering is involved, but with the additional advice from ACGM taken into account.

FRANCE

In France, the approach to biotechnology has been to rely as far as possible on industry-sponsored voluntary standards to supplement existing legislation. France's source of specialist advice on experiments involving recombinant DNA is the Commission Nationale des Recombinaisons Génétiques (CNRG), which was formed in 1979 to cover rDNA research. This has also evolved to provide advice on larger-scale work with industrial applications, but referral to the Commission by industry remains voluntary. Formal guidelines for large scale use of genetically engineered organisms are being considered by the regulatory authorities and private industry. Pharmaceutical products are controlled in the same way as their non-genetically engineered counterparts, with the exception that there is a voluntary requirement to remove DNA from the final product. Further guidance is under consideration by the Laboratoire Nationale de la Santé, but at present a flexible case-by-case review is undertaken when new drugs are considered for approval under existing legislation.

In the field of deliberate release, there is a similar lack of specific regulations and guidelines, and any releases of genetically engineered organisms could be considered only under existing environmental controls to prevent pollution. Controls would apply if the organisms had a pesticidal property, through the Agricultural Chemicals Control Law of 1943. However, there are no controls at present on research or field tests involving biological pesticides.

GERMANY (FDR)

In the Federal Republic of Germany, the equivalent to the RAC is the Control Commission for Biological Safety (CCBS), whose membership structure is more analogous to the UK's ACGM, since it includes worker representatives as well as scientific experts. Control outside of the area of government-funded research has been voluntary, but steps were taken in 1986 towards requiring the registration of all laboratories carrying out work involving genetic engineering, and the prior approval of experiments using infectious organisms or toxins. Products of genetic engineering are regulated in the same way as their non-engineered counterparts, although there has also been a need to obtain a special permit to culture in excess of 10 litres of a genetically engineered organism. Pharmaceuticals are covered under the Law on the Reform of Drug Legislation and there are no plans to amend this for the time being.

Existing laws also cover the introduction of agricultural products, including biological pesticides. New products must be registered under the Plant Protection Act of 1978 before their introduction on to the market, and the rDNA guidelines currently expressly prohibit the deliberate release of genetically engineered organisms. Foods and food additives are also regulated under general law and this would provide a basis for the regulation of any genetically engineered foods or food additives, such as single-cell protein.

More general environmental laws may also be applicable to the products of biotechnology. For instance, the Federal Emission Law of 1974 may be applicable to the airborne discharge of micro-organisms, the Water Resources Act prohibits the discharge of living micro-organisms, the rDNA guidelines make specific provisions requiring the destruction of all living material before release into the environment. As these stand, therefore, it would not be possible to test genetically engineered organisms in the environment. A committee of enquiry was set up in 1974 to investigate gene technology, and this reported late in 1986, and included a call for a 5-year moratorium on any deliberate releases of genetically engineered organisms, reinforcing the current prohibition on such experiments.

EUROPEAN COMMISSION

The UK, France and West Germany are all members of the 12-nation European Economic Community (EEC) which has as one of its primary goals the removal of restrictions to trade within the EEC. The disparity of different national laws and regulations on the products of biotechnology could well delay the introduction and spread of new products, and the EEC has therefore taken a number of steps to encourage the harmonisation of national regulations.

In the area of pharmaceuticals, the EEC has set up a system to simplify the approval of products throughout the community. Council Directive 65/65/EEC set up the basic regulatory framework for market approval of human therapeutic products, Directive 75/318/EEC established common requirements for testing and evaluation of medicinal products, and Directive 75/319/EEC set up the Proprietary Medicinal Products Committee (CPMP) to allow a company with full clearance in one member state to apply for approval in other states. Recent amendments now require review by the CPMP only if the authorities in the other countries object to the application within 60 days.

Harmonisation is well advanced on the notification of new chemicals under the Sixth Amendment (79/831/EEC) of 1979 to the Council Directive on Dangerous Substances. This requires the premarket notification of all new chemicals and the provision of a minimum set of data on the substance and its properties. This has led to uniform procedures and data requirements throughout the EEC, but this is unlikely to be applicable to genetically engineered organisms, and it appears unlikely that the Commission will attempt to expand its definition of chemical substances to include organisms along the lines of the approach taken by the US Environmental Protection Agency.

In other areas, EEC harmonisation has not progressed as far, and the regulation of pesticides, food and food additives is still essentially left to the different systems of each member state.

The environmental applications of biotechnology have been considered in some depth by the Commission and proposals have been drafted for a Community-centred assessment of any proposals. Similar proposals for the centralised consideration by CPMP of new drugs (before the national authority) have also emerged, but it is not yet clear if and when these will result in an EEC-based approval scheme, and it may be some time before companies can regard the 12 countries of the EEC as a single regulatory unit.

JAPAN

Japanese guidelines in rDNA research were finalised in 1979 and were generally considered to be more stringent than the NIH RAC guidelines. They are applied to universities and government research agencies by the Department of Education and elsewhere by the Science and Technology Agency. Relaxations in the guidelines were, however, implemented in 1983 and 1986 with specific provisions to reduce restrictions on large-scale fermentation (over 20 litres) by industry. A number of organisms which have been confirmed as safe by the laboratory can now be mass-cultured by industry without prior approval. This covers, for instance, interferon, insulin, growth hormones expressed in E. Coli, or yeasts. Other organisms requiring higher containment levels still require prior authority. The Ministry of Industry and Trade (MITI) has jurisdiction over large-scale industrial applications, while other departments regulate the products.

Pharmaceutical products are approved by the Ministry of Health and Welfare under the Pharmaceutical Affairs Act of 1983. Several sets of standards may apply, including the Japan Antibiotic Drug Standards, the Biological Preparations Standards, the Radiopharmaceutical Standards and the 'Preparation of Data Required for Approval Applications for Drugs Manufactured by the Application of Recombinant DNA Technology'. Further guidance may be published on other aspects involving rDNA, such as the facilities and equipment used in the manufacture of rDNA drugs, mammalian cell culture, food products and quality standards.

The Ministry of Education has also put out guidelines on the use of recombinant DNA in experiments involving plants and animals. Physical isolation of genetically engineered plants is required, and a guideline for the deliberate release of rDNA plants is being developed. At present, the general guidelines on rDNA prohibit the deliberate release of recombinant organisms into the environment.

A number of issues are currently relevant. Firstly, there is general concern at the impact that regulatory delays and difficulties may have on the emerging biotechnology industry, particularly if international approaches to regulation differ markedly. Secondly, the issue of deliberate release of genetically engineered organisms into the environment is a consistent source of controversy. Thirdly, the possibility of export controls on biotechnology may become a significant issue in the future. Finally, many governments remain concerned to assist their biotechnology industry to bring products to the market as soon as possible. The remainder of this chapter will discuss these four issues in more detail.

7.3 Current Issues

INTERNATIONAL COORDINATION OF BIOTECHNOLOGY REGULATIONS

Mention has already been made of the moves within the European Community to harmonise regulatory approaches to biotechnology within the twelve member states. A broader effort has also been undertaken within the 24-nation Organisation for Economic Cooperation and Development (OECD), which brings together not only Europe, but also North America, Australasia and Japan.

The OECD decided to review Recombinant DNA Safety Considerations at the 1983 meeting of the Scientific and Technological Policy Committee, and a special working group of experts from member countries was set up. This reviewed the positions of each country on the safe use of genetically engineered organisms at the industrial, agricultural and environmental levels, and identified what criteria could be developed for authorisation of their use. The group's report and recommendations were accepted by the OECD Council on 30 May 1986 and published. This provides a single set of general considerations and guidance which may help member countries to reduce the disparities between national regulatory regimes.

The report examined the safety considerations associated with large-scale industrial use, and for environmental and agricultural applications. It concluded that the vast majority of industrial rDNA applications will use organisms of intrinsically low risk which warrant only minimal containment (Good Industrial Large-Scale Practice). The use of organisms of higher risk necessitates the use of additional criteria for risk assessment and may require additional physical containment. Such techniques are well known to industry and have been successfully used for many years. There was thus general endorsement for continuation of the regulatory approaches developed for the traditional biotechnology industry.

ENVIRONMENTAL RELEASE

In the case of industrial fermentation, the genetically modified organism is contained within the manufacturing plant and usually destroyed by sterilisation before it is released. In the case of genetically engineered organisms intended for release into the environment, however, additional risk factors must also be considered, including the possibility of ecological disruption, effects on non-target organisms (infectivity, pathogenicity), and the possibility that genetic material might be transferred to other organisms (e.g. the transmission of herbicide resistance to weeds).

The OECD group recognised that the assessment of the potential risks of organisms for environmental or agricultural use was less well developed than for industrial applications, but considered that the methods for assessing the consequences of releasing rDNA organisms could be approached by analogy with the extensive data bases gained from the use of traditionally modified organisms in agriculture and the environment generally. It concluded that by a step-by-step assessment during the research and development process, the potential risk to the environment from rDNA organisms could be minimised.

The most commonly envisaged releases of genetically engineered organisms into the environment are in the agricultural area – primarily with respect to genetically engineered plants and micro-organisms which act as plant pest controllers or as plant growth stimulators or protectants. Examples already known at the research stage include plants with engineered pesticide resistance, viruses which destroy serious pests such as caterpillars, soil bacteria containing a pest toxin gene, and bacteria with the gene for ice nucleation removed. Other applications which may be beneficial in a less direct manner include inoculants for increasing the rate of decomposition of compost or silage, and organisms for speeding the degradation rate of straw.

Environmental applications of microbes are also commonplace – particularly in waste water treatment. There is also interest in the possible use of micro-organisms to degrade toxic substances in the soil. Indeed, the first large-scale cleanup of a contaminated land site employing optimised (though not genetically engineered) bacteria has just been successfully completed in the UK, where bacteria mixed into the contaminated soil succeeded in degrading phenols, cyanides and sulphides in an abandoned gas-works site to below target levels within a year. No proposals have been made to date for any such environmental applications of genetically engineered organisms, but the possibility exists that performance could be enhanced by this technology.

We have already seen that differences are emerging in the national approaches to release. Some countries already have regulations prohibiting the release of genetically engineered organisms; others have neither a specific prohibition nor approval mechanism, while some have established procedures for evaluating and approving proposals. The UK and the USA are the furthest advanced in this area, since both have conducted trials of genetically engineered organisms under the relevant approvals.

In the UK, the ACGM considered their first application for the planned release of a genetically manipulated organism, and approved the proposed experiment, in 1986. The experiment involved the use of a virus to infect caterpillar pests, which had been used for some time in its natural form to combat outbreaks of the pine beauty moth in Scotland. The genetic engineering merely inserted a non-coding marker sequence into the genome, so that the survival rates and spread patterns of the virus could be determined. Also in the UK, a transformed potato plant was grown in the field, outside contained facilities, in 1984. This experiment used the bacterium *Agrobacterium rhizogenes* to infect wounds in potato cultivars, resulting in the transfer of genetic material from the bacterium to the potato plant. Expression of the introduced genes caused stable alterations in plant development and tuber shape which were retained under field conditions.

In the USA, approval was given by the RAC in 1983 for a proposal to test in the field a bacterium *(Pseudomonas syringae)* with the ice-nucleation gene removed. This also received an Experimental Use Permit from the EPA under FIFRA in 1985. This case has, however, become involved in extensive litigation by the Foundation for Economic Trends and has also been re-examined by the EPA when it was revealed that unauthorised uncontained experiments had taken place before the statutory approval was given. Indeed, despite many projects having been put forward to the RAC or EPA, only 2 'deliberate releases' had taken place by 1986. The first was the approval and marketing of a genetically engineered vaccine for pig scours (approved by the USDA); the second involved a genetically engineered tobacco plant with disease resistance incorporated.

The problems encountered through litigation (and most recently through civil liability complications) over proposals for the deliberate release of genetically engineered microbes have persuaded some companies to delay or abandon projects. On the other hand, the genetic engineering of plants has not excited quite the same response, since it is easier to reassure the public that the experiment can be contained and the plants destroyed if necessary.

This erratic progress is partly a result of the public and pressure group concern over the issue. The spectre is raised of genetically engineered pests 'out of control' on the lines of rabbits in Australia, Dutch elm disease in the UK, Kudzu vine and gypsy moth in the USA. At the same time the continuous application of traditional genetic manipulation in agriculture over many years, and the fact that rDNA offers more precise well understood means of effecting genetic change, are not readily communicated to nor understood by the public and legislators. The prohibition of deliberate release in Denmark and, more recently, in the Netherlands (which has an otherwise very active record in supporting biotechnology) shows that this is likely to remain a difficult and contentious issue, which will necessitate care and sensitivity on the part of scientists and regulatory authorities if public concerns are to be met without stifling these applications of biotechnology.

EXPORT CONTROLS

The USA has in recent years led a very active campaign to restrict the availability of high technology goods to the Eastern bloc and some other countries. For this purpose, the Export Administration Act has been widely used not only to restrict the activities of companies in the USA, but also to apply extraterritorially to any company using a US product. In the Federal Register publication of the Coordinated Framework, the Export Administration Act was stated to apply to a very wide range of substances related to biotechnology, including viruses, yeasts, bacteria and recombinant DNA.

Up until now, these powers have not been applied with the same vigour as they have for computers, materials and the like. However, it is known that the Defense Department and the Department of Commerce have been considering a more focused list of biotechnological materials which might be subject to export controls. This issue could emerge to restrict trade in biotechnology, and is one which the industry will wish to monitor.

ASSISTANCE FOR SMALL BIOTECHNOLOGY COMPANIES

Many of the companies involved in biotechnology are well established with their own regulatory affairs departments familiar with the needs for different countries. Biotechnology has, however, also spawned a large number of smaller companies which are attempting to generate products in highly regulated markets – particularly pharmaceuticals, but also agricultural and environmental applications. Many of these companies lack not only the experience in dealing with different countries' regulatory systems, but also the financial resources to withstand unforeseen delays in the process of getting a product through the developmental and regulatory approval stages and to the market. Consequently, some governments have attempted to ease the way for such companies.

Some regulatory authorities have set up special offices designed to assist small manufacturers, such as the FDA's Office of Small Manufacturers' Assistance (OSMA). Others have published relatively simple guides to their regulatory systems (e.g. the UK's 'Plain Man's Guide to Support and Regulations for Biotechnology'). Some governments also offer advice on other countries' regulatory systems as part of export assistance packages – either in the relevant department in the country concerned or locally in the embassy or consulate in the country to which the product is being exported. Both may provide not only information but also potential consultants who might be employed to represent company interests abroad.

Other sources of assistance or guidance can be found in industry associations, and their international links (e.g. the UK's Association for the Advancement of British Biotechnology and the US's Association of Biotechnology Companies have a reciprocal membership agreement which would allow members of one to call on advice from the other association). Such sources of advice can be invaluable for the new company with limited resources and experience.

7.4 Conclusions

The biotechnology industry has moved rapidly from the research stage to the development and marketing of products. Many of the early hypothetical hazards attributed to working with rDNA in the laboratory have not materialised, and the scientific community now generally regards genetic engineering as only one of many techniques available for use in the microbiology laboratory. As such, it can be governed within the existing framework of good safety practices for biological laboratories.

In the large-scale industrial uses of recombinant DNA technology, a consensus has emerged within the OECD that organisms of intrinsically low risk can and should be used wherever possible under the conditions of Good Industrial Large-Scale Practice, which have developed over the years from experience with a range of micro-organisms. This can be expected to meet most of industry's needs, since most hosts of recombinant genes have been expressed in safe, well-characterised hosts. Methods of physical containment are available if it is necessary to handle organisms requiring higher containment levels. The availability of such international guidance may assist international harmonisation of national regulations on this aspect.

On the other hand, the issue of deliberate release of genetically engineered organisms is proving highly controversial and has led to considerable litigation in the USA, with a number of countries having placed severe restrictions on experiments in the field. The issue has become something of a 'Green' pressure group issue in some countries, and great care will be needed by regulatory authorities to proceed cautiously and with full public disclosure of the objective facts of each case. Major delays in the testing of some genetically engineered organisms have already occurred and some companies have withdrawn projects in this area. There is a possibility therefore that these regulatory uncertainties will be seriously detrimental to the development of agricultural and environmental applications of biotechnology.

8 Chapter Eight
Patenting In Biotechnology

8.1 Introduction

Over most of the last two decades, the subject of patent protection in biotechnology has been beset by three major questions and related issues. First, there has been the debate over the patentability of micro-organisms themselves (as distinct from processes for producing or utilising micro-organisms). Secondly, there has been the question of how broad a patent can be in respect of either the micro-organisms themselves or the accompanying process technology and the resulting products. This second question has yet to reach its conclusion in the field of genetic engineering inventions, in which it may turn out to be the most problematic issue of all. Thirdly, as regards the contribution of the inventor, there have been major case law developments concerning the adequacy of the patent disclosure (the specification) for microbiological inventions based on new micro-organisms and their uses.

Interest in these questions has not been confined to the patent lawyers. Businessmen and scientists, academic as well as industrial research workers, and economists also have seen patent protection as an important element in the advancement of this technology. This is evidenced strongly by the special review of international patent issues made by the Organisation for Economic Cooperation and Development in its multi-part investigation of all facets of biotechnology. (See 'Patents' section in Information Resources chapter, for references and legal precedents quoted in this chapter.)

Many of the important patentability issues may in future be focused most sharply upon the areas of recombinant DNA inventions and inventions in the field of cell fusion such as in hybridoma and monoclonal antibody technology. Indeed, this phase has already opened with the beginnings of litigation in the United States and in Europe so that we can look forward to the establishment of precedents in case law for these technologies. Nevertheless, some of these issues are still not completely settled in relation to patenting in the field of classical biotechnology, and the important developments that have taken place in that area should first be recalled as background to an appreciation of the problems of patentability in the new biotechnology.

8.2 General Issues in Patenting

PATENT PROCEDURE

The time scale of patent procedure is illustrated in Figure 1 below. An application for patent protection is normally first made in the country of residence or place of business of the applicant. This establishes a so-called 'priority date' which will be recognised in most of the other countries of the world under the provisions of an international agreement known as the Paris Convention. In practice, this means that the major expense of a foreign patenting programme can be postponed until towards the end of one year after the initial filing date in the home country. For this purpose, an application for a European (regional) patent is on the same footing as national applications in other countries filed under the Paris Convention. The value of this one-year interim period, both to industry and to other organisations which have the problem of assessing the potential industrial importance of new research results, is considerable.

The other major advantage given by the Paris Convention is that the inventor can publish details of his invention at any time after his priority date without detriment to his patent prospects. The only provisions here are that the invention is clearly defined and well supported by data in the first application and that the foreign applications are filed no later than one year after the first application.

As indicated in Figure 1, the patent application is published in some countries while still at the stage of being an unexamined application (18 months after the priority date). This is the norm in Europe under the EPC and most European national laws (also Japan), but in the USA and Canada publication does not occur

158

Figure 1   Patenting procedure.

Time   Invention

0   1st application, establishing priority

1 year   2nd application, consolidating protection in home country

                        and

                        Foreign applications (with priority)

                        Official prior art search

18 months   Publications of application (some countries)

Variable   Official examination
period   and prosecution
(typically
3 years
or more)

                        Grant or refusal of patent

until the patent is granted. The specification can be amended during examination and the final claims may be of narrower scope than first drafted.

PATENT CLAIMS

Patent claims are expressed in the most broad and general language which the patent attorney can devise to avoid loopholes which competitors can exploit. The claims must be adequately justified by the experimental data provided by the inventor. A typical process claim will specify the starting materials and the products of the process and the reagents and experimental conditions necessary to convert the one to the other. In the microbiological process claim, the main centre of interest is the definition or description of the micro-organism itself. The product claim is one which defines the product itself. The best form of product claim is known as the product *per se* claim and this gives what is commonly described as absolute protection for the product itself irrespective of how it is made.

In the product *per se* claim, the entity must be defined by its constitution, by what it is rather than by its method of manufacture ('product-by-process') and not simply by what it does, i.e. its properties. The definition of products for claiming purposes has been fairly straightforward for relatively small molecules but is more difficult for high molecular weight substances such as synthetic polymers and enzymes. In a rapidly moving technology, patenting cannot usually wait for the elucidation of full chemical structures, and it has been necessary to use other parameters to obtain product claims for large molecules. These can be based on combinations of chemical data, spectral data, physical constants, and biological properties which together can be said to define the product uniquely or with as much precision as science will allow. These techniques of claim drafting have been used successfully in the patenting of enzymes and other natural products, but have been of little use for the patenting of living matter.

Before discussing the latter problem in detail, it will be helpful to recapitulate the main categories of invention in microbiology established before the advent of genetic engineering inventions.

Patentable Microbiological Invention
1. Process of producing a new micro-organism.
2. The new micro-organism as produced by the defined process.
3. The new micro-organism *per se*.
4. Process of cultivating or otherwise using a known or a new micro-organism to produce an end-product which may be:
    (a) a form of the multiplied micro-organism itself; for example vaccine or edible biomass;
    (b) a by-product of microbial growth, for example an antibiotic, enzyme, toxin, or an otherwise useful industrial product (even if inactive biologically);
    or
    (c) some other product or substrate which is produced or

improved by the culturing process, for example a purified industrial product or effluent.

5. The product of any of the processes defined in (4) – defined by a *per se* claim or product-by-process claim as appropriate.

6. Particular formulations of defined strains or cultures thereof, including combinations with other substances, designed to utilise and exploit their special properties, for example in human or animal foods or for industrial uses.

BASIC CRITERIA OF PATENTABILITY

To obtain patents of any kind it is necessary that the subject matter covered by the claims meets patent law criteria of novelty, inventiveness, and utility or industrial applicability. In addition, the supporting description of the invention must be adequate for a person of ordinary skill in the art concerned, e.g. another scientist, to reproduce the process or product described and claimed.

In the patent law of most countries, the novelty condition requires that the invention must not already be available to others by any kind of public disclosure or public use before the date of filing of the patent application. This condition is described as 'absolute novelty' because it covers disclosure anywhere in the world and by any person whatsoever including the inventor himself who may destroy his chances of obtaining patent protection by public disclosure in the scientific literature or elsewhere (say, at a conference) before taking the first protective step of filing a patent application. The United States, Canada, Japan and a few other countries are exceptions to this rule of absolute novelty in that they allow grace periods in which publication or use by the inventor will not prejudice a later filing for patent protection so long as it takes place before the grace period has expired. The first major genetic engineering patent, the well-known Cohen-Boyer patent, was obtained by making use of the US grace period of one year after publication.

In addition to being novel the invention must also not be obvious to the ordinary skilled worker from previously recorded knowledge (the state of the art), i.e. it must not follow plainly or logically from what is already known. Furthermore, the invention must have a practical application; under US law this is expressed as a 'utility', whereas under European law the corresponding requirement is to be capable of 'industrial application'. The US concept of patent utility is much wider than industrial utility and can cover all sorts of methods which are useful outside a strictly industrial context.

At the moment, research workers and others in biotechnology seem to be most interested in the possibility of product patents for micro-organisms *per se*.

If the micro-organism is a known strain available to the skilled worker it is sufficient to refer to it by name. But for a *new* micro-organism, the skilled person requires not only a description but also the means of access to the micro-organism. Legally, the identification of the micro-organism is necessary to clarify the scope of the patent but from the practical standpoint access to it is also essential to provide an 'enabling disclosure'. This problem was first appreciated in antibiotics technology and was solved by depositing new strains in culture collections from which they could be made available to readers of patent specifications. The maxim that 'what cannot be described must be deposited' has since become enshrined in patent law either by court decision or by statute.

8.3   Patenting in Classical Microbial Biotechnology

Patent statutes do not single out particular technologies for attention except where legal policy excludes specific items from patentability such as computer programs, plant and animal varieties, and the actual diagnosis or treatment of disease, these being denied patent protection under European law. But most recent patent laws and regulations have made specific provision for what are described as 'microbiological inventions' and these will be outlined later. A microbiological invention is presumably one involving a micro-organism, but patent law has so far shown no inclination to define this term, thus allowing itself flexibility to embrace inventions involving other biological entities which replicate or can be made to replicate, e.g. plasmids, viruses and animal or plant cell lines. The practical rules established for microbiological inventions have been extended to these other entities without any apparent difficulty.

The prototype of the microbiological invention is the cultivation of a new strain of micro-organism to produce a new product, or to produce a known product in a better way under specific fermentation conditions. Process technology of this kind has

162

been handled effectively under patent law by building upon the experience of a century or more of chemical process patenting. Indeed, chemical patenting as a whole has provided the backcloth against which biotechnology inventions are first viewed by the patent lawyer and administrator. Chemical inventions have given us four categories: process, product, use, formulation or composition, into which microbiological inventions fit fairly comfortably. These categories of invention correspond to the patent claims, a claim being a verbal formula defining the essentials of an invention and usually also the legal scope of protection conferred by the patent.

THE DEPOSITION OF MICRO-ORGANISMS FOR PATENT PURPOSES

The practice of making so-called 'patent deposits' of microorganisms first arose for inventions involving new strains isolated from nature or from mutation programmes. It is now so embedded in the thinking of many Patent Office officials that it continues to be raised in connection with new strains produced by genetic manipulation in spite of the fact that these techniques are more amenable to reproduction from a written description.

The first legal decision on deposition was given in the United States of America in the celebrated Argoudelis case in which the application claimed two new antibiotics and a new microbiological process for their production. The process involved a new *Streptomyces* strain which had been isolated from nature and deposited, before the filing of the US patent application, with a public depository in the USA. The micro-organism had been deposited under the condition that access to it would be restricted to persons authorised by the applicant in the period before the patent was granted. The US Patent Office had argued that this deposit was secret and confidential and therefore that the application was defective because the micro-organism had not been made available to the general public at the time of filing of the patent application. The court held that it was not necessary for the public to have access to the culture before issuance of the patent. The Argoudelis case was therefore chiefly concerned with the most sensitive element of deposition practice, i.e. the date on which the deposited strain must be made freely available to competitors and other members of the public.

This particular issue has been even more troublesome under European patent law, as will be explained later. For many years

163

after the Argoudelis decision it was assumed that the deposit would have to be made with the culture collection no later than the filing date of the patent application so that the latter could cross-refer to the deposited strain by its accession number. More recently this assumption has been modified in a US court decision (Lundak case – see Part 3) on a case where through oversight the patent application was made some days before the strain was deposited. The inventor, a university professor, argued that his cell line was effectively on deposit in the university laboratories and elsewhere and would have been available to the Patent Office during the pendency of the application and to the public after grant as a result of the formal deposit with a recognised national culture collection. This argument succeeded.

Although this helpful decision is welcome, it is not in line with current European patent regulations on depositions, and adds to the existing lack of harmony in international patent practice in this field. In view of this decision, the US Patent and Trademark Office is currently revising its guidelines and policy on deposition procedure.

If deposition provides a legal solution to the descriptive problem, it does not necessarily provide a commercially acceptable solution. Deposition is required not merely to provide a reference material for Patent Office purposes, but more importantly to enable samples to be made available to third parties. The main controversy on this issue is over the date of public release of the culture and the conditions that should be attached to this release. Under US patent law, the situation, though not ideal, is generally acceptable whereas under the corresponding laws in European countries the timing and other conditions of release of samples of deposited micro-organisms are far from satisfactory to industry and other applicants for patents and are seen to disfavour the patent route as compared with the alternative of industrial secrecy.

The European Patent Convention (EPC) was set up in 1973 and came into operation in 1978 providing a regional patent system for European contracting states. This provides for a single patent application processed in the European Patent Office which when granted matures into a 'bundle' of separate national patents in any of the states designated by the applicant when the European patent application is filed. At present there are 13 contracting states in which protection can be achieved by this convenient and cost-effective route (Austria, Belgium, France, Germany (West),

Greece, Italy, Liechtenstein, Luxembourg, Netherlands, Spain, Sweden, Switzerland, United Kingdom).

AVAILABILITY OF DEPOSITED ORGANISMS

We have seen that, under US law, availability of the deposited strain is not required before the grant of a patent. In Europe, under the EPC, availability is covered by Rule 28 of the regulations which form part of the Paris Convention. Rule 28 requires an applicant to:
  i. deposit the new organism no later than the European patent application date
  ii quote deposit accession data in the application
  iii accept that a sample of the deposited strain will be made available from the date of first publication to either (a) any person or (b) an independent expert nominated from an official list by a third party. (The independent expert, though commissioned by a third party, must not pass the strain on to his principal.)
Alternative (i) will apply unless the applicant opts for alternative (ii) before preparations for publication of the application are complete. Availability under either (i) or (ii) is subject to certain undertakings, namely:
1) While the patent application is pending or so long as the eventual patent lasts, the culture cannot be passed on to others by the person obtaining the sample from the culture collection.
2) While the application is pending (but not after grant of the patent), the person obtaining the sample undertakes to use the culture for experimental purposes only.
Both conditions cease upon refusal or withdrawal of the patent application.

Rule 28 applies not only to the deposited micro-organism but also to cultures derived from the deposited organism and still exhibiting the characteristics of the deposited organism essential to carrying out the invention. This must include sub-cultures of the deposited strain and modifications thereof which meet the terms of the definition. Finally, it should be noted that the Rule applies only to micro-organisms not already available to the public 'and which cannot be described in the European patent application in such a manner as to enable the invention to be carried out by a person skilled in the art'. This wording allows the possibility of avoiding the need to deposit the new micro-organism where the applicant can justifiably rely on the reproducibility of the described method of producing and identifying the micro-organism, e.g. by a repeatable technique of genetic manipulation.

The present situation on the release of deposited strains is summarised in Figure 2. UK and Germany are examples of countries which do not allow the independent expert any option for national applications filed in their own countries, i.e. non-EPC applications, whereas the current EPC practice has been adopted by France and Sweden for national applications also.

Figure 2   Release of deposited strains.

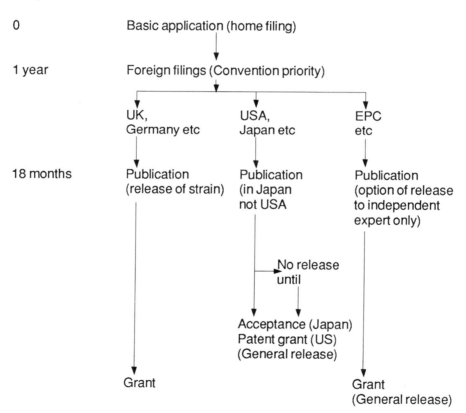

Industry and other interested circles desire harmonisation of these varying practices in a way favourable to applicants by allowing for greater control over the precious new strain in the period before an enforceable right is obtained. The US and

Japanese positions represent the most favourable scenario, the present UK and German the worst. European Patent Office practice is a compromise, but still far from ideal. Even after the grant of a patent there should be some restrictions on the freedom of one who has obtained the deposited strain, e.g. a prohibition on export to a territory not covered by an equivalent patent. More can be done to remove the present disincentive to utilise the patent system for inventions of this kind.

THE BUDAPEST TREATY

This is an international convention governing the recognition of deposits in officially approved culture collections for the purposes of patent applications in any country that is a party to it. By the beginning of 1986, nineteen states had signed this convention. According to the Budapest Treaty a culture collection may become designated as an International Depositary Authority (IDA) and thereby become recognised by all the contracting states. A single such depositary, whatever its location, can therefore be chosen by an applicant to hold a deposit relevant to a single patent application or to a family of related patent applications filed in any number of contracting states. An IDA must store a deposited micro-organism for at least 5 years after the most recent request for a sample and for at least 30 years from the original date of deposit.

8.4 Micro-organism Patents: Case Histories

Most important countries now allow patent claims for micro-organisms *per se* subject to certain reservations. First, there is the distinction between newly constructed micro-organisms, i.e. man-made, and those which are new in the sense that they have been isolated from nature for the first time. Attempts to patent the latter meet with the objection that the organism is a 'product of nature' and must therefore be unpatentable. Whilst this attitude is not found everywhere in such extreme form, it is the underlying reason for the many restrictions which are placed upon claims of this kind.

In 1980 the United States Supreme Court allowed a patent to A. Chakrabarty with the following claim:

'A bacterium from the genus *Pseudomon as* containing therein at least two stable energy-generating plasmids, each of said plasmids providing a separate hydrocarbon degradative pathway.'

This bacterium was a manipulated oil-degrading organism but not obtained by recombinant DNA technology. The court held that this claim was not to a hitherto unknown natural phenomenon but to a non-naturally occurring manufacture or composition of matter, a product of human ingenuity having a distinctive known character and use. In also rejecting the product of nature argument the court held that the patentee had produced a new bacterium with markedly different characteristics from any found in nature and one having the potential for significant utility. This was not nature's handiwork but that of the inventor.

The Chakrabarty case was clearly influential in persuading the Canadian Patent Office (1962) to allow a claim to a combination of five micro-organisms (some new, some old) acclimatised to spent sulphite liquor. The European Patent Office also allows claims to micro-organisms and even though this practice has not been confirmed judicially there is no reason to question its validity.

Without doubting the seminal importance of the Chakrabarty decision it is worth noting that the first high-level judicial decisions on the patentability of micro-organisms have come from Europe. In Germany, in the so-called Baker's Yeast Case (1975), the applicants claimed two mutant strains of yeast by reference to their specific culture collection accession numbers. The German Supreme Court held that in order to obtain protection for a micro-organism *per se* (including one found in nature), the organism must be capable of being obtained by disclosure of a repeatable method of producing it without dependence on biological material provided by the applicant (such as by propagation from a sample deposited by the applicant). In the patent specification no method was disclosed for producing the mutants other than through propagation of a culture of these same mutants. This was sufficient to meet the requirement of the prevailing patent law.

Also in Germany, in the 7-chloro-6-demethyltetracyclin case (1977), the invention was concerned with the production of mutants of a type strain of *Streptomyces aureofaciens* which would selectively produce the title compound in preference to related tetracyclines. The German Supreme Court rejected the claims on the grounds that the mutants were obtained by treatment of the

type strain with ultra-violet radiation or chemical mutagens but
no details of the process had been given which would have been
reproducible with reasonable certainty of success and with the
expenditure of reasonable effort and cost.

In the *Lactobacillus bavaricus* case (1978), however, the German
Federal Patent Court upheld a claim to a group of naturally
occurring micro-organisms to which the applicants gave the
species name *Lactobacillus bavaricus*. The claim defined the micro-
organisms as 'obtainable' by carrying out certain specified selection steps which resulted in the production of bacteria which
predominantly produced the L(+) isomer of lactic acid. Although
naturally occurring, the new micro-organisms had previously
been undiscovered and required human technical intervention to
recognise them and produce them in a reproducible manner. The
subject matter of the application was therefore an invention and
not a mere discovery. The court was persuaded that a reproducible description enabling the skilled person to produce (i.e. isolate)
strains of this species had been given in the specification.
Germany is the most prominent of a small group of countries that
take this view of the reproducibility requirement. Most others
accept the deposit system as overcoming the problem.

NATURALLY OCCURRING MICRO-ORGANISMS

To patent organisms of this type one must overcome the product
of nature objection mentioned above and also the related
objection that such organisms are mere discoveries and not
inventions. In the Ranks Hovis McDougall case (1976), the claims
rejected by the Australian Commissioner of Patents were of the
following type:

'*Fusarium graminearum* Schwabe deposited with the Common-
wealth Mycological Institute and assigned the number IMI
14525 and variants and mutants thereof.'

These novel strains had been developed as suitable for the
production of edible protein. They were admittedly isolated from
soil samples, but no disclosure of the particular soil or particular
method of isolation had been given. The specification was held
insufficient for this reason. Availability through a culture collection was not enough. The view was also taken that as the organism
was isolated from soil, 'There may at best have been a discovery.
No invention was involved in the mere discovery or the mere
identification or the mere isolation by an unspecified method of

something that occurs in nature.' In asking the rhetorical questions 'What has the inventor done, what contribution has he made?' the reply was 'He has discovered a naturally occurring micro-organism and by altering its conditions of growth he has changed its morphological characteristics. If that is all that he has done he has made no useful contribution to the art.'

At the extreme negative end of the spectrum of opinion, therefore, the isolation of 'found' micro-organisms is regarded as mere discovery and therefore unpatentable. The middle ground taken by most commentators is that the technical teaching of an isolation procedure or the mere fact of one having been required adds the element of inventiveness necessary for patentability. It is assumed that it was not obvious (a) that the specific organism existed, (b) how to isolate it and (c) that it would have the useful properties shown for it.

The adherents of this position usually suggest that the claim to the micro-organisms must not be for the organism in its natural state but must include some kind of purity limitation as, for example, in the USA where the claim must be for a 'biologically pure' culture. The most positive approach to patents in this situation regards the purity limitation as contrived and artificial. For what use was it to the skilled worker that the micro-organisms existed in nature if that existence was previously unrecognized and therefore unutilised? In practice, such a restricted claim may give sufficient protection in most cases, but one can imagine circumstances in which the purity limitation may be inconvenient and impose a difficult burden of proof on the patentee.

The 'product of nature' question continues to be a point of difficulty in patent law. In the older case law, the question arose in connection with the isolation of relatively simple naturally occurring compounds, e.g. adrenalin, Vitamin B_{12} and certain steroids. In its modern context, it re-surfaces in relation to the patentability of natural proteins purified and isolated for the first time in quantity from cell culture systems or from genetically programmed organisms. It is also involved in the intriguing question of the patentability of natural gene sequences. Questions of this type are now common in the field of the new biotechnology.

8.5 Patenting in the New Biotechnology

Many of the categories listed earlier for classical biotechnology will have their counterparts in the newer techniques of micro-organism construction such as by gene splicing into plasmids and other vectors or by cell fusion to produce hybridomas and other new animal cell lines. In plant genetic manipulation there will be corresponding possibilities of protection. Twelve areas for consideration are:

Patentable inventions in genetic manipulation
1. Novel genetic engineering strategies.
2. Isolation of genes.
3. Modification of genes.
4. Synthesis of natural or modified genes.
5. Construction of gene inserts.
6. Vector systems.
7. Methods of transforming cells.
8. Transgenic organisms.
9. Manipulating plant protoplasts and the like.
10. Methods of regenerating whole plants.
11. The new plants.
12. Transgenic animals.

In the development of patent practice in this field, the major question will be not about the categories as such but about the scope of patent claims allowable. Already there are indications, particularly in Europe and Japan, that patents may be rather narrow, e.g. limited to specific DNA sequences or specifically described and deposited plasmids and transformed cells.

PATENTS FOR GENETIC MANIPULATION

The Cohen-Boyer patent was the first patent granted in the United States for recombinant DNA techniques, and many patents in this field have since been issued. Claims to synthetic genes and plasmids in terms of functional properties and specific restriction maps as well as those drafted more concretely in terms of DNA sequences are all appearing, together with claims to recombinant strains and other scientific creations. It will be the case law which decides how broad patents can be in this technology. The greatest interest will focus around the level of inventiveness required to sustain claims of the broad scope that are required to stop off effectively the loopholes through which competitors can slip with their own specific sequences and clones.

On the question of inventiveness, the Wistar Institute case (1983), decided by the British Patent Office, is particularly noteworthy. This involved the application of cell fusion techniques to the construction of hybridomas for the production of monoclonal antibodies. After the original discovery of the basic technique and the subsequent appreciation of its general importance, it was reasonable to assume that the patentability of any application of this general procedure must rest on some special and non-obvious property or advantage of the particular system constructed, in much the same way as should apply to the patenting of particular applications of recombinant DNA technology. Mere novelty should not be enough.

In the Wistar Institute case the claims were so broadly drafted as to cover the application of the basic method to the preparation of any antibodies to viruses. These were refused. Rejection of an attempt to re-patent the basic technique as applied to such a broad sub-class of antibodies is perhaps not surprising. However, the Patent Office also rejected a specific claim to an individual novel hybridoma described in the Wistar application on the grounds of lack of any stated exceptional quality which would justify patent protection.

In the USA a liberal approach was taken by the Court of Appeals for the Federal Circuit in sustaining the Hybritech patent for an immunometric assay system using monoclonal antibodies (*Hybritech Inc.* v. *Monoclonal Antibodies Inc.*). The invention lay in the use of monoclonal antibodies in place of prior art polyclonal antibodies in a known sandwich assay system. When the US Patent Office originally examined the application it argued that it would be obvious to use monoclonal antibodies in place of polyclonals in conventional immunoassay protocols.

This objection was overcome by including in the claims a numerical limitation regarding the affinity (binding power) of the antibodies. In the District Court the patent was held invalid on the grounds of obviousness, as many would have predicted, but this was overruled by the remarkable decision of the higher court. Much of the case turned on aspects of US patent law relating to priority of invention which have no parallel in the patent laws of Europe and most other countries. Reference to laboratory notebooks and other evidence in order to determine priority of invention is crucial under the US first-to-invent system, but is inadmissible under the European or any other first-to-file system of patent law. But on the question of inventiveness the court upheld the patent because the prior art was 'devoid of any suggestion that monoclonal antibodies can be used in the same fashion as polyclonals'. Also, it was influenced by the commercial success of the patentee's product and found that a 3-year time gap between the first availability of the monoclonals and the sale of Hybritech kits was long enough to indicate lack of obviousness. The court also gave short shrift to the 'obvious to try' argument. This decision can be seen as part of the general tide now flowing in favour of patents, which is certainly welcome. But we still await from litigation the guide to what is reasonable in terms of claim scope for inventions in the field of genetic engineering.

Recombinant DNA patents are beginning to show a certain pattern. The specification usually begins by describing the nature and activity of the protein to be produced and referring to the shortcomings of previous methods of extraction, from which the protein can be produced only in small quantity. Then follows a glossary of genetic engineering terms and a series of drawings illustrating the applicant's cloning strategy, including the plasmids to be used for this purpose and their restriction enzyme sites and the appropriate DNA sequences for insertion therein. The specification then describes the detailed methodology in densely written experimental protocols with frequent reference to publications in the scientific literature. For the non-expert it is hard going to plough through this material, and one can appreciate the problem of the Patent Office examiner who must evaluate the sufficiency of the disclosure and pinpoint the inventive step.

The claims which follow are usually worded very broadly in functional terms rather than being limited to the specific details of the process. The European Patent Office is now hardening its practice against claims which simply define products in functional terms. For example, a claim to a microbiologically prepared protein having a certain biological activity is not allowable simply on the basis of its preparation by genetic engineering techniques. Similarly, any claim to 'a nucleotide sequence coding for' a certain peptide will be allowed in broad terms if the peptide is new, but if the peptide is of known structure the sequence must be defined by specific nucleotides. These claims will be more easily allowed if they refer to material in vectors or host organisms deposited in culture collections.

The Biogen European Patent for the production of alpha-interferon by these techniques was opposed by 9 companies and ran into difficulty with the announcement by the Opposition Panel that it intended to revoke the patent. The Biogen specification, while admitting the use of generally known technology, presented its case by stressing the difficulty of locating and separating DNA sequences of unknown structure which code for the expression of alpha-interferon.

The question of claim scope will also be crucial to the outcome of the current British case involving Genentech's UK patent on tissue plasminogen activator (TPA) produced by recombinant DNA methods. Is the first research group to produce a particular protein by genetic engineering methods entitled to broad product protection which in effect covers all such ways of making the same product or a product so similar in composition that it is essentially the same biological activity? In the TPA patent the invention is said to be 'based on the discovery that recombinant DNA technology can be used successfully to produce human tissue plasminogen activator in amounts sufficient to initiate and conduct animal and clinical testing as pre-requisites to market approval'. One of its product claims reads 'Human tissue plasminogen activator as produced by recombinant DNA technology'. The issue for the court in this case seems clearly drawn.

Reverting to US litigation, the Hormone Research Foundation patent covering methods of producing synthetic human pituitary growth hormone (hGH) is said to be infringed by a Genentech product. The patent has a product claim to a synthetic product of defined amino acid structure shown in the drawings. This structure, which is said to be the true structure of the natural hormone hGH, differs from the structure previously assigned to the hormone in the scientific literature. This claim raises the interesting question whether it is possible to patent the synthetic form of a natural product in the circumstances outlined.

INVENTIONS IN PLANT SCIENCE AND TECHNOLOGY

Plant cell and tissue culture methods and the use of plant cells, e.g. in immobilised or other form, for the production of valuable chemical products, are considered by patent examining authorities to be within the general category of microbiological process inventions. These are accordingly patentable as processes and, up to a point, the new products of such processes may also be patented. The point at which difficulty is encountered is when the attempt is made to patent a new plant as such. In Europe this difficulty arises from the European Patent Convention which in Article 53(b) states that European patents shall not be granted in respect of:

'plant or animal varieties or essentially biological processes for the production of plants or animals; this provision does not apply to microbiological processes or the products thereof.'

The question arises: are new plants produced by genetic manipulation to be considered as 'plant varieties' and therefore unpatentable as products? This writer suggests an interpretation which would exclude patents for plants bred by traditional methods but not those produced by microbiological methods. This raises important and controversial issues which impinge upon the other legal system for protection, namely, that of plant variety rights (plant breeders' rights, certificates of variety protection, etc.). The latter form of protection, which is limited to commercial production and sale of reproductive material, is generally considered less satisfactory than patent protection.

At present, in the attempt to patent genetic manipulation techniques applied to plants, it seems to be possible to claim the methods and intermediate products up to and including the stage of transformed plant cells. However, claims to plants regenerated from such cells meet with objection under Article 53(b).

This official vigilance in Europe against attempts to patent plants has even been applied to a European patent application (Ciba Geigy Application, 1984), claiming chemically treated propagating material, i.e. seed treated with certain oxime derivatives in order to confer resistance to agricultural chemicals. Fortunately, the Appeal Board overruled this objection and thus threw some light on this rather obscure part of the law.

In US patent law, no such difficulties arise from the written statute. Indeed, some important patents have been granted in the past for procedures that would be classed as essentially biological in European terms, e.g. well-known patents for hybrid maize based on utilising cytoplasmic or genetic male sterility. More recently, the US Patent Office attempted to refuse a patent on a maize seed having a specified minimum endogenous free tryptophan content achieved by tissue culture and selection techniques on the ground that this was only protectable under the Plant Variety Act of 1970. However, the Patent Office was overruled on appeal. The US law remains, therefore, the most open and flexible system for protection in plant biotechnology.

Representatives of the international plant variety right system (UPOV) have already expressed uneasiness over the possible consequences of patents covering genetic material which is inserted into the plant genome. Of particular concern is the extent to which the owner of a patent for a DNA sequence, for example, could exert his rights through to the finished plant variety and whatever further commercial use or research activity might be made of it. This debate will intensify as plant genetic manipulation becomes closer to commercial reality.

8.6 Conclusions

Biotechnology has challenged and strained the patent system in ways that it has not experienced from other technologies. The response which courts and legislators will have to make during the next few years will determine whether, after all, patent law can fulfil a role in stimulating and encouraging investment in this field. The task of comprehension of this technology by lawyers is alone formidable enough.

Added to this, the problem of harmonisation of laws is also in need of urgent solution. Technology is already moving ahead into the areas of the higher life forms with the genetic manipulation of plants and animals. Patent law will need to adapt swiftly to the pace of technological advance if it is to continue to be an attractive option to the inventor and the investor.

9 | Chapter Nine
Impact on Medicine

9.1 Introduction

Life is an irreversible process – finite and very precious. The markets for products which preserve or improve life and health are consequently large and profitable. They are also self-promoting: to paraphrase Albert Einstein, 'As the circle of health increases, so does the circumference of illness.'

During the past three decades man has developed tools to help him look at life analytically and is beginning to understand that modulation. Furthermore, man can synthesize the building blocks of life inside large-scale laboratories. The application of analysis and synthesis is the story of the impact of biotechnology on medicine.

9.2 Production of Human Proteins

Major advances have been made in understanding how our genetic material, DNA, directs the production of the proteins which form the structure of, and which regulate the processes in, our bodies. As mentioned in earlier chapters, the manipulation of DNA in a deliberate and controlled fashion has become known as genetic engineering, which basically consists of taking a gene from a chromosome of one type of organism and inserting it into the chromosome of another. This permits biotechnologists to dissect out individual components of complex living systems, and to mass produce them in relatively simple micro-organisms or mammalian cells inside fermentation tanks.

Using genetic engineering for the production of human proteins depends on (a) being able to isolate the DNA interest,

(b) selecting a suitable organism in which to insert and produce it, and (c) being able to extract and purify the product after fermentation. The most frequently used producer organisms are the bacteria, *E. coli* and *Bacillus subtilis*, the yeast *Saccharomyces cerevisiae*, and various mammalian cell lines which can be grown in bulk culture. The earliest genetic engineering work used *E. coli* because it was the micro-organism which was best understood at the genetic level and because very high concentrations of protein can be produced in the bacterium. As a production host for proteins to be used in medicine, however, it has two major disadvantages – firstly, certain strains are known to cause intestinal infections in children and, secondly, the proteins that it produces are usually retained within the cell, which can complicate extraction and purification processes.

B. subtilis, on the other hand, is not known to cause disease and it can excrete proteins. The trouble with *B. subtilis* and with other bacterial hosts, is that the proteins they produce are not exactly like those which are produced by the human body: natural human proteins have polymers of sugars added to them after they are synthesized (glycosylation). For this reason, and, perhaps more importantly because they have been used for centuries in the production of human foodstuffs and beverages and are considered safe, yeast cells have also been used as production hosts. However, the sugar polymers which yeasts add to cloned proteins are also different to those on the natural proteins. Mammalian cell cultures, therefore, are increasingly being used for the production of proteins from recombinant DNA. Many of the mammalian cell lines which produce useful amounts of proteins are derived, however, from tumours, a fact which can complicate the requirements imposed by regulatory authorities on protein production processes. Each of these production hosts has advantages and disadvantages, and which one is eventually used for a particular product will be determined by clinical studies, the attitudes of the regulatory authorities and the quantity of the protein demanded by the market. The various human proteins currently being produced by genetic engineering can broadly be classified into three categories: hormones, blood products and lymphokines.

HORMONES

Many hormones are relatively simple protein molecules which, in the body, provide organ-to-organ communication. The best known example is insulin, a protein essential for correct sugar metabolism, which is produced by the pancreas and is lacking in diabetics. The action of most of those which have been cloned is relatively well understood. In 1983, human insulin produced in *E. coli*, marketed as Humulin by Eli Lilly, became the first commercially significant clinical product produced by recombinant DNA. Its performance in the market has, perhaps, been disappointing: preliminary studies had not shown any significant benefit of human insulin over the insulins from animal sources which already satisfied the diabetic market. To make Humulin competitive, Eli Lilly raised the price of their porcine and bovine insulins, and while this may have gained Humulin an entry into the US market, it also allowed in the Danish company, Novo, who had produced their own version of human insulin by making a small biological change to conventional pig insulin. In 1987, Novo have announced a continous process for the production of genetically engineered human insulin in yeast cells, a method which they claim simplifies purification of the product.

The market for human growth hormone (hGH) is like that for insulin to the extent that this product already existed before recombinant techniques came along. hGH extracted from the brain tissue of human cadavers has been used for some time as a treatment for dwarfism. But in the spring of 1986, the US government banned the use of pituitary hGH following the death of four patients from Creutzfeldt-Jakob syndrome, a rare virus disease which could have arisen from the brain extracts. Britain, Greece, Sweden, the Netherlands and Belgium soon followed suit, opening the way for the recombinant product (Protrophin) marketed by Genentech which had been granted US marketing approval in October 1985. In the first financial half of 1986, Genentach's revenues from Protrophin (their first from direct sales of any recombinant product) were $16.8 million, and the worldwide market is just opening up. KabiVitrum of Sweden, one of the companies whose natural hGH products were withdrawn following the Creutzfeldt-Jakob scare, received approval in the UK and Belgium for its recombinant hGH (Somatonorm) developed under contract by Genentech.

Despite the relatively small size (in pharmaceutical terms) of the world market for hGH ($100 million a year currently), numerous companies are involved: in the USA, the major players joining Genentech are Serono (whose development work has been performed by Celltech in the UK), Eli Lilly and California Biotechnology, while in Europe, Sanofi, Nordisk and Biotechnology General as well as KabiVitrum are active. Biotechnology General (BTG) and Serono/Celltech are developing methods for the production of hGH identical to the natural hormone (Protrophin and Somatonorm both have an extra amino acid group being made in bacteria); BTG has isolated an enzyme which cleaves off the amino acid group, while Serono/Celltech are using mammalian cell culture methods. The intrinsic purity of the recombinant hormones and the relative ease of producing large quantities of hGH may mean that, by the time the safety question concerning the natural hormone has been resolved, there may be no market left for it.

Calcitonin is a thyroid hormone involved in regulating calcium retention in bones. In post-menopausal women and in patients suffering from Paget's disease, calcitonin levels are often reduced and bones become structurally weakened. The high cost of hospitalising patients with fractures which result from this weakening means that the calcitonin market could currently be worth in the region of $200 million. However, as yet there seems to be little commercial activity. At Celltech, calcitonin and a closely related hormone, katacalcin, have been cloned in bacteria, and the company is collaborating with Sankyo to exploit these findings. Sandoz in Basel and Unigene in New Jersey are independently investigating CGRP (calcitonin gene-related peptide): Sandoz has filed a patent on cloned CGRP and believes that there may be a market for the peptide in the treatment of cardiovascular disease.

The other major market for hormones is in the treatment of pain. If you hit your thumb with a hammer, it is a natural reaction to shake your hand vigorously or rub the affected part hard. Shaking or rubbing helps release natural painkillers in the body, the endorphins. These are small molecules which bind to the same pain receptors to which traditionally painkillers like morphine bind. Since the endorphins are quite small molecules, the role of biotechnology in their production may be limited; they can probably be produced more efficiently by chemical means. However, genetic engineering and molecular biology will make significant contributions to understanding the pain receptors which they bind. By studing the interactions between the endorphins and their receptors, it is hoped that painkilling drugs without addictive effects can be designed.

180

BLOOD PRODUCTS

Around 50 of the proteins found in human blood have now been cloned in *E. coli* or mammalian cells, and yet only a few of these are being used in clinical studies. The most advanced studies are being performed with tissue plasminogen activator (tPA), a large enzyme which dissolves the blood clots which cause thrombosis and heart attacks. Other thrombolytic agents, streptokinase and urokinase, are already on the market but in contrast to tPA, when dissolving blood clots they can also cause substantial internal bleeding. Genentech are again the leaders in this field and their tPA has successfully come through clinical trials. (US marketing approval is expected soon.)

A feature of tPA development, and one which reflects the huge potential markets for this drug (estimated at between $200–900 million by 1990), is the extensive competition characterised by the close involvement of major pharmaceutical and chemical companies with genetic engineering firms. Of the European majors, Germany's Boehringer Ingelheim are Genentech's partners, BASF are linked with Integrated Genetics, Hoechst with Chiron, Sandoz of Austria with Collaborative Research, and Wellcome (UK) with Genetics Institute. Of the European biotechnology companies, Celltech are tied in with Sankyo, and Biogen with SmithKline in the USA and Fujisawa in Japan. Porton International in the UK is also producing recombinant tPA, and Bio-Response and Damon Biotech are seeking pharmaceutical partners for cell culture derived products. In most of these cases, tPA has entered or is about to enter clinical trials.

Work is also well advanced in the production of genetically engineered Factor VIII as a treatment for haemophilia. The market here contrasts sharply with that of tPA and is perhaps more typical of the markets for blood products. There is only a small number of haemophiliacs worldwide (and hence a much smaller market), but each of them would require repeated injections of Factor VIII throughout their lives. In contrast, only short courses of tPA would be required for each patient. Preparations, therefore, would have to be extremely pure so that side effects and allergies did not develop. Clinical research has been given impetus by worries about AIDS. For many years, Factor VIII has been extracted, along with other blood products, from donations of human blood. The risk of AIDS and hepatitis from this source (several haemophiliacs in the UK contracted the disease from preparations of Factor VIII imported from the USA) has led to a reappraisal of the genetically engineered variety. Other blood products which became commercial include Antithrombin III and Factor XIII.

LYMPHOKINES

The lymphokines are a group of small proteins (including the interferons, interleukins, and tumour necrosis factor) which are hormone-like in action but which exert their efforts through cell-to-cell interactions rather than over the whole body. This is an important distinction with relevance to their use as pharmaceuticals, since most drugs are currently administered systemically.

The cloning of alpha interferon in *E. coli* in 1980 was heralded as a breakthrough for biotechnology and fuelled investment interest considerably. Since then, however, progress has been steady but less than spectacular. Regulatory approval has been granted to Schering-Plough (collaborating with Biogen) and Hoffmann-La Roche (collaborating with Genentech) in several countries for the use of alpha-interferon against relatively rare conditions such as hairy-cell leukaemia and Kaposi's sarcoma (an AIDS-associated cancer), and promising results have been obtained in clinical trials involving other cancers. The basic problem with alpha-interferon, and one which may be common to many similar biological response modifiers, is that it appears to have a vast range of actions in the body, some of which result in serious side effects. Many cancer patients complained of nausea and pain, and while alpha-interferon in nasal spray form could be shown to restrict the spread of the common cold virus, patients suffered all the normal cold symptom – aches, runny nose and inflammation – they were caused by interferon!

Gamma-interferon is expected to make a much greater impact than alpha interferon. Biogen's product has already received approval in West Germany for the treatment of rheumatoid arthritis, and Shionogi are conducting RA trials on Biogen's behalf in Japan. As a product of the immune system, gamma-interferon is expected to have some impact in the treatment of autoimmune diseases like systemic lupus erythromatosus and multiple sclerosis. Interleukin-2 has also entered cancer trials, with Cetus in the USA, Biotest Serum Institut in Germany and Ono Pharmaceutical in Japan among the leading companies.

For biotechnology companies, the large size of potential

markets for compounds which can combat cancer, combined with the ability to produce large amounts of these compounds through genetic engineering, has possibly clouded the fact that little is known about cancer and less about the actions of the compounds against it. It is unclear how the potential of these biologicals can be best exploited, and trials are now beginning on combination therapies. For instance, Hayashibara in Japan have promising results from trials in which a combination of alpha-interferon and tumour necrosis factor (TNF) have been used against breast cancer, and Sandoz have combined Genetics Institute's colony stimulating factor (CSF) with chemotherapy and radiotherapy in trials against cancer, AIDS and aplastic anaemia.

In many cases, therefore, the products of recombinant DNA technology have yet to find their markets. As the complexity of the interactions between the lymphokines is unravelled by fundamental studies in biotechnology, a more rational approach to the design of cocktails of anticancer agents (both recombinant DNA and conventional) may develop. It is clear now that there is probably no single 'wonder drug' which will cure all cancers. The biological changes leading to cancer are now better understood, but the situation has been revealed to be very complex. Biotechnology, particularly genetic engineering and cell culture, has been a key tool in such studies.

9.3   Vaccines

Vaccination has eradicated smallpox and has drastically reduced the incidence of diphtheria, tetanus, tuberculosis, polio, cholera and whooping cough in developed countries. Many infections remain, however, as challenges for biotechnology: virus diseases such as influenza, hepatitis and AIDS; parasitic diseases like malaria, filariasis, toxoplasmosis and Leishmaniasis; bacterial diseases such as shigellosis and leprosy.

A vaccine is basically a sheep in wolf's clothing – a living or non-living preparation which can fool the body's immune system into thinking that it is a pathogenic organism. The body has no way of distinguishing a harmless invader from a pathogen; it can only determine that it is harbouring foreign material (an antigen). The presence of the antigen initiates a complex sequence of actions, the aim of which is to contain, neutralise and destroy the invader. One of these actions is the production by the body of antibodies which bind specifically to the antigen. Antigenic material also stimulates the immune 'memory' allowing the body to act more quickly and effectively in combating subsequent invasion by the same organism or, importantly, by organisms which, to the body's immune system, appear similar.

An effective vaccine should have the ability to stimulate the immune system without causing any symptoms of disease. Furthermore, it should not itself cause adverse effects in the body.

Biotechnology provides several ways of 'dissecting out' the antigenic material from its disease-causing background. One is to use recombinant DNA techniques to extract from the pathogen a gene (DNA) which is responsible for the production of a protein which is known to be important in stimulating the immune response. The gene is then placed into cells of another type of organism (which might be another micro-organism or a mammalian cell) which can be grown in large amounts in bioreactors and induced to produce the protein. If this antigen protein can be extracted and purified, it could be used as a form of 'killed' vaccine.

For instance, scientists at the USA's National Cancer Institute in collaboration with RepliGen, Cambridge, MA, have used the intestinal bacterium, *Escherichia coli*, to produce small protein fragments derived from a surface protein of the AIDS virus. When these fragments are injected into goats, the goats produce antibodies which neutralise test tube preparations of the AIDS virus. Similarly, researchers at Genentech in San Francisco, CA have produced the whole AIDS virus surface protein in a mammalian cell line which can be grown in bulk culture. This preparation has been shown to protect experimental animals against certain strains of AIDS virus but not others. These vaccines incorporate just a single protein antigen and hence are safer than conventional vaccines which contain (hopefully only) dead cells or a weakened strain which might very occasionally revert to its original virulence.

Recombinant DNA methods can also produce new 'live' vaccines. Again using an AIDS example, the gene for the AIDS surface protein has been inserted into a harmless virus called vaccinia which can grow to a limited extent in the human body without causing disease. (Vaccinia was originally used to eradicate smallpox.) In the hybrid, the AIDS protein is present on the surface of the virus and thus potentially available to stimulate the immune system.

The third approach to new vaccines is the use of synthetic chemical methods to mass produce small pieces of antigenic proteins. For this approach, the three-dimensional structure of the antigenic protein must be determined in order to indicate which particular regions might be expected to be involved in immune stimulation. The most advanced work in this field has been performed with the viruses which cause influenza. The 'flu virus is able to alter its external appearance so that the immune system is ineffective in protecting the body against subsequent invasion. This leads to regular epidemics of 'flu worldwide. By studying several different types of virus, scientists hope to be able to identify small regions of the external protein which are both essential to virus function and invariant, and could thus form the basis of a vaccine which will be effective against all strains.

The fourth contribution of biotechnology to vaccine production is the development of anti-idiotype antibodies. These are antibodies which mimic the three-dimensional structure of antigens and which, when administered, can induce an immune reaction against the original antigen. The way in which they are produced is a bit like casting a mould. Imagine the original antigen as a stone statuette; the immune reaction against the antigen produces antibodies the structure of which specifically reflects the 3-D structure of the antigen, like a layer of plaster of Paris on the statuette. These specific antibodies can then be used to direct the production of antibodies, the anti-idiotype antibodies, which mimic the original antigen in the same way as the plaster mould can be used to produce virtually identical statuettes from, say, bronze. The anti-idiotype antibodies can then be used to induce antibodies which will cross-react with the original antigen. The advantage of the anti-idiotype antibody approach is that it should be completely safe. No pathogenic antigens are ever presented to the body. However, no such vaccines have yet been licensed.

The production of purified proteins via recombinant DNA methods is the most commercialised area for new vaccine technology with the most advanced market being vaccines against hepatitis B, a highly infectious disease which is also implicated in liver cancer with 200 million carriers worldwide and a potential vaccine market by 2000 AD of $250 million. Scientists at Biogen were the first to extract a piece of DNA which codes for a surface antigen from the hepatitis virus. They inserted it into the chromosome of yeast and produced complexes of the hepatitis B surface antigen protein (HbSAg) which can be purified and used as vaccines. The vaccine is currently in human trials in Japan and Europe.

A similar approach was adopted at Genentech where HbSAg was expressed in mammalian cells and used successfully in chimpanzees: Mitsubishi Chemical Industries are licensing Genentech's technology and hope to commercialise the vaccine by 1990. In fact, both Biogen and Genentech have been beaten to the market by Chiron Corp. Their recombinant DNA Hepatitis B vaccine has obtained approval in West Germany and the USA where it has been licensed to Merck, Sharp and Dohme who have spent $2.5 million on a facility for the production of 600,000 doses per annum and forecast sales of $40–45 million during 1987. Pasteur Vaccins are expected to launch a similar product soon, while the products from Endotronics and Integrated Genetics will begin clinical trials in 1987.

Malaria is another important disease for which recombinant vaccines are being developed. The causative organism, a parasite called *Plasmodium falciparum*, takes many different forms during its complicated life cycle in an effort to elude the immune system. For instance, it enters the body in a form known as a sporozoite which rapidly penetrates the liver, effectively disappearing from the clutches of the immune system. It has been found, however, that the sporozoite surface is covered by a single protein antigen and the genes coding for this have now been cloned and produced in *E. coli*. The resultant vaccine entered phase II efficacy trials during 1986 and the World Health Organisation (WHO) are currently planning large-scale phase III trials. SmithKline Beckman, one of the companies producing the vaccine, forecast that it will be 25–30 per cent cheaper than conventional vaccines against the disease.

Other potential disease markets for recombinant vaccines include herpes, rabies and cytomegalovirus (which often causes fatality in newborn infants). A further interesting example is the development by Biotechnology Australia of a vaccine against a hormone called inhibin, to treat infertility.

9.4 Monoclonal antibodies

Antigen recognition by antibodies, which is so important in the body's defence system against disease, has also been exploited for many years in the development of diagnostic tests for diseases using antibodies from animals. In 1975, the pioneering studies of Cesar Milstein and George Kohler at the Medical Research Council Laboratory of Molecular Biology in Cambridge, UK, led to the development of hybridoma technology. Antibody-producing cells were fused with tumour cells to produce a hybrid which could be grown in laboratory culture and which reproduces antibodies of a single specificity, monoclonal antibodies. The commercial potential of monoclonal antibodies was soon obvious, the technique had not been patented, and numerous small companies were established to exploit the technology. Monoclonal antibodies represent one area where biotechnological products have been sold for some time, and it is more realistic to look at the markets than to examine individual companies or products in detail.

Celltech in the UK produced some of the earliest products sold – monoclonal antibodies which could be used to purify recombinant DNA products like interferon from the culture broths in which they were being produced. Originally research tools, these monoclonals, now produced in bulk, are playing an important role in helping the therapeutic proteins to reach their markets.

MABS IN DIAGNOSTICS

Monoclonal antibodies are also being increasingly incorporated into health-care diagnostics. In 1985, over 70 monoclonal diagnostic products had been approved by the US Food and Drug Administration (FDA) and represented sales of $30–50 million per year. This number is continuously increasing and one report suggests that by 1990, the world monoclonal diagnostic market will be worth $1.1 billion, or around 12.5 per cent of the total market for in vitro diagnostics. Another report predicts that, in Europe by 1991, at least half of all immunoassay diagnostics (a $1.4 billion market) will contain at least one monoclonal component.

The largest suppliers of monoclonal-based reagents to the European markets have been two US companies, Hybritech and Monoclonal Antibodies, but Biotest Serum of West Germany and Sorin Biomedica of Italy have each claimed over 10 per cent of the total sales. Biotest have achieved notable success in marketing monoclonal ABO reagents in Germany, as have Celltech in the UK. However, Amersham International of the UK and Abbott

Laboratories of the USA remain the two leading suppliers in the wider immunoassay market, and the rate at which large companies like these incorporate monoclonal reagents in their tests will be a major governing factor in the growth of monoclonal diagnostic technology.

Another important consideration in the development of markets for monoclonal diagnostics is the competition from the newer field of DNA probes. Monoclonals will not be used in tests for susceptibility to genetic disease or for drug resistance determinations in infectious micro-organisms. They do, however, have a great advantage in areas like therapeutic drug monitoring and cancer diagnosis where DNA-based tests are impossible or complex. For all other areas (largely infectious disease testing) where direct competition with DNA is likely, monoclonal antibodies have at least one big advantage: diagnostic personnel are very familiar with immune-based techniques, and the equipment for performing immunoassays is already in place in their laboratories. DNA probe tests use skills more akin to genetic engineering and some require quite complex procedures.

The innate ability of monoclonal antibodies to bind specifically to a particular substance is the basis of their utility. The extent of their use in diagnostics, however, will also be determined by the methods by which that binding is converted into a measurable signal, the label. At present, most immunoassays (monoclonal or otherwise) use radioactive antibodies (radioimmunoassays or RIA): if the sample is positive, the antibody will bind and the radioactivity will be retained on the sample. The stringency of rules governing the use of radioisotopes means that immunoassays using this approach are largely restricted to centralised specialist diagnostic facilities.

Several other labelling techniques now exist and are growing in importance. In enzyme immunoassays (EIA), a colour-producing enzyme is coupled to the antibody and results can be read either by eye or spectrophotometrically. An interesting variant of this technique is the enzymic cascade developed by IQ(BIO) in the UK in which several enzymic reactions are coupled to produce a significant amplification of the original binding signal. Fluorescence immunoassays (FIA) and luminescence immunoassays (LIA) are related techniques in which the label emits fluorescence or light, respectively. All these systems are achieving increasing penetration into existing immunoassay markets and, furthermore, are creating a new market – assays which can be performed in doctors' offices or surgeries or even in the home. It is in this area of immunoassay that monoclonals are expected to have the greatest penetration. Monoclonal Antibodies have already produced two diagnostic dipstick tests for home use (for pregnancy and ovulation) and these are likely to be followed by diagnostics for sexually transmitted diseases and other infections.

MABS IN IMAGING AND THERAPY

Perhaps the greatest problem in the treatment of cancer is that cancerous cells are indentical to normal cells in almost every respect. Therapeutic agents which will kill cancer cells, therefore, are very likely also to kill normal cells. It has been found, however, that the surfaces of cancer cells in the body differ in a few respects from those of normal cells. Since monoclonal antibodies recognise specific antigens on cells, they are being used at a clinical research level to image tumours and in therapy against melanomas, lymphomas, breast and colon cancer.

Several different approaches are being investigated: antibodies can be used either alone or linked to toxic drugs. Centocor, a US company which has recently established a subsidiary in the Netherlands, are collaborating with Hoffmann-La Roche in Switzerland and are using monoclonals alone against gastrointestinal cancer. The theory is that when the antibodies bind to the tumour, they attract the cells of the immune system to act against the cancerous tissue. Centocor are using this approach initially because it allows the company to go into clinical trials faster.

It is believed, however, that naked monoclonals will not, in most cases, be effective in cancer therapy. One reason is that some cancer cells are able to swallow up the antibodies which bind so that they are able to attract immune cells. However, this means that monoclonal antibodies may usefully be used to target toxic drugs to the interior of cancer cells. Eli Lilly in the USA have linked a compound called desacetylvinblastine (which incidently is a product derived from plants and could be produced by plant cell culture) to a monoclonal that recognises an antigen on lung, breast, prostate and pancreas cancer cells. The conjugate has been shown to be effective against human tumours implanted in mice. Similarly, Farmitalia Carbo Erba in Italy are working with Cytogen of New Jersey on the development of monoclonal conjugates of anthrocycline drugs.

A further refinement of this approach is to use monoclonals to target extremely toxic compounds, such as ricin from the castor bean and the toxin from the bacteria which causes diphtheria, to tumour cells. There are obvious problems in using these toxins – the targeting must be extremely specific to ensure that the compounds do not enter normal tissues, and the toxins are relatively large molecules which might destabilise the conjugate, but several companies are nevertheless pursuing this approach. Cetus and Xoma are probably the leaders in the USA, and the UK's Imperial Cancer Research Laboratory is also involved. The same monoclonals which are being used to target drugs are, in many instances, also being used to image tumours by conjugating them to radioactive elements. Following injection of the conjugates, body scanning techniques can be used to localise and quantify cancerous tissue, enabling clinicians to give an initial diagnosis or to assess whether the disease is responding to conventional treatments. These *in vivo* diagnostic methods represent a large market in themselves and provide much needed income for companies waiting for therapeutic agents to emerge from clinical trials.

9.5 DNA Probes and RFLP Analysis

DNA PROBES

Monoclonal antibody diagnostics exploit the ability of antibodies to lock on to specific biological components by virtue of their three-dimensional complementarity. In contrast, diagnostics based on DNA probes exploit two-dimensional complementarity. DNA is a double stranded linear polymer strand composed of many thousands of monomers. There are only four different monomer types, the nucleotides known as A, T, C and G. The two strands are held together by weak forces between pairs of these monomers: A always binds with T, and C with G. When DNA is heated, these weak forces are disrupted resulting in the formation of two single strands of DNA; on cooling, the specificity of the interaction between A–T and G–C pairs means that native, double-stranded DNA molecules can reform. This specific pairing is the basis of DNA probe diagnostics.

To produce a diagnostic kit for a venereal disease like gonorrhoea, for instance, short pieces of single-stranded DNA (the probe) are produced, either by recombinant DNA techniques or by chemical synthesis, which will specifically bind to a known sequence of the DNA of the gonococcus. Clinical samples can be treated to produce single strands of DNA which can then be mixed with the probe. Binding which occurs between the sample DNA and the probe can be detected by labelling the probe either with radioactivity or with an enzyme which produces a coloured product.

Much of the current work on DNA probes centres on the development work of labelling systems to make such diagnostic tests more sensitive. In contrast to diagnostics based on antibodies, where there may be many sites on the surface of an organism where the antibody can bind, there is usually only one site for the binding of a DNA probe. One solution being developed by Enzo Biochem of New York is to attach chains of a molecule called biotin on to their DNA probes. Each biotin molecule can bind a colour-producing enzyme, thereby creating a greatly increased signal. At Cetus, the problem has been confronted head on: the company is using DNA probes to direct the amplification of the relevant DNA sequence within clinical samples. By increasing the amount of DNA, its detection becomes easier.

The FDA approval for the use of DNA probes on patients' specimens was recently given to Gen-Probe of San Diego, CA for diagnostics for two bacterial infections, walking pneumonia and Legionnaire's disease, and it is anticipated that this will signal the rapid development of a DNA probe market worth a billion dollars by 1990. Infectious diseases will form an important part of this market: by the end of 1987 Cetus, in collaboration with Eastman Kodak, aim to launch their test for AIDS into a world market worth $100 million a year; Enzo Biochem have entered the research market with tests for herpes I and II and *Chlamydia*, and are planning to enter the AIDS market; Integrated Genetics are conducting field trials with cytomegalovirus probes. In the study of cancer, DNA probes are likely to play an important role in helping to understand the course and causes of the disease. Probe diagnostics for cancer, however, are probably some way off. The techniques will have to be refined so that they can not only indicate the presence of a particular DNA sequence (a cancer-causing gene, or oncogene), but can also pinpoint its exact location within the human chromosomes.

However, the largest markets for DNA probes may lie outside both the infectious disease market and cancer – in the diagnosis of genetic diseases such as Duchenne muscular dystrophy, cystic fibrosis, sickle cell anaemia and Huntington's chorea. Because DNA probes home in on the genetic material itself, they are ideally suited for locating defects in human DNA which are associated with heritable disease.

RFLP ANALYSIS

Every human being inherits one set of genes from its mother and another from its father. In most cases the sets of genes from both parents are identical. Diseases like haemophilia and Duchenne muscular dystrophy can be traced back to mutations of genes we inherit from one or other of our parents. As well as these obvious differences between the two sets of DNA we inherit, there are many thousands of 'silent' differences known as restriction length fragment polymorphisms, RFLPs for short.

If a woman were a carrier for an X-linked genetic disease such as Duchenne muscular dystropy, it would mean that there was a genetic defect on one of her X-chromosomes. The law of averages would indicate that this defect would have a 50 per cent chance of being passed on to any children she might have; a 50 per cent chance, therefore, of passing on the diseased state to her sons or the carrier state to her daughters. Any pregnancy in a family with a history of X-linked disease was fraught with dilemma. For DMD, in which muscle wasting begins at around the age of three followed by death in the mid-teens, it was usual to terminate all pregnancies involving male foetuses despite the fact that half of them would have led to the birth of normal children.

However, it is now possible to diagnose DMD and other genetic defects prenatally. Cells of the growing foetus can be obtained by techniques such as amniocentesis or chorionic villi sampling, and DNA preparations made from them. Enzymes which cut DNA in specific places are used to trace 'silent' genetic markers, the RFLPs, associated with the disease. By comparing the pattern of these markers on the foetal DNA with the patterns on the maternal and paternal DNA it can be determined with a high probability whether the foetus is normal or diseased. Probe systems are being developed for a number of different diseases which are believed to be due to single genetic defects: Duchenne muscular dystrophy, cystic fibrosis (CF), polycystic kidney, sickle cell anaemia, Huntington's chorea. In addition, researchers are starting to look at DNA probe diagnosis for conditions like heart disease, schizophrenia and Alzheimer's disease (an age-associated neurological disorder). Most heavily involved are two companies located in Massachusetts, Collaborative Research and Integrated Genetics. The tests currently available are hardly routine, but genetic diagnosis is being offered as a laboratory service at around $1000 per diagnosis.

This high cost reflects the fact that diagnosis currently involves tracking the genes through a family using genetic markers which are near the defect of interest. Eventually it is hoped researchers will be able to close in on the genetic defects themselves using these distant markers as a starting point. Then DNA probes will be able to be used directly to detect genetic differences between normal and diseased states, making for a much faster and cheaper diagnostic test.

9.6 Conclusion

The ability to locate defective genes with DNA probes is obviously the first step towards developing treatments for these defects. However, until treatments are available, the ability to diagnose without the ability to treat raises several ethical issues. For instance, Huntington's chorea is a severe degenerative disease which does not strike until the mid-twenties or later. In a recnt survey of people at risk for Huntington's disease, 96 per cent thought that genetic tests should be made available and 66 per cent said they would want to be tested. While the patient's right to know his or her fate is indisputable, the psychological consequences of such revelations are as yet uncharted.

The question also arises: should employers have the right to use genetic screening as part of the hiring process? While it could have obvious benefits in, for instance, preventing the exposure of susceptible individuals to harmful chemicals in the workplace, it might also allow employers to discriminate against individuals who were genetically prone to heart disease or cancer. And even if employers were prevented from using genetic screening, would prospective employees be obliged to reveal the findings of any screening performed in the course of their normal health care programme? Similarly, genetic screening could mean that certain individuals find themselves unable to obtain suitable life and health cover insurance. This kind of ethical question is now also being extended to the rights of employers and insurers to test for infectious diseases (primarily AIDS) and for drug abuse. This is raising many questions of confidentiality and personal liberty and promises to have a significant social impact in the next two decades.

Meanwhile, at a more academic but no less important level, a vast project whose aim is to analyse the whole of the human genome using recombinant DNA techniques has recently been begun in the United States. While the information this generates will undoubtedly enhance our understanding of human health and disease, it will also be open to abuse. Human potential is not defined merely by a genetic blueprint, and it is vital to the health of society that human ethics and morals develop in tandem with technology.

The role of biotechnology is to transform fundamental knowledge of biological systems into useful medical practice. In taking science to society, what is useful must be defined in terms not only of what can be done but also what is morally acceptable and commercially justified (though these may not always coincide). These further requirements considerably complicate any assessment of the likely impact of biotechnology on medicine, but it is clear from developments to date that there will be significant effects from biotechnology not only on the tools of medicine but on the very way in which it is practised.

10 Chapter Ten
Market Trends and Prospects

10.1 Introduction

1987 is poised to be the year that biotechnology finally begins to prove itself in the marketplace as well as the research laboratory. Introduction of new biotechnology-produced pharmaceuticals could more than double US industry revenues to about $1 billion in 1987, and regulatory approvals for several similar products are pending in various European countries. Japan's bioindustry market, which current estimates put at about $300 million, will double by 1990, with pharmaceuticals and food applications driving growth (see Chapter Five). It is the 1990s, however, that will demonstrate biotechnology's real market clout: current estimates are for industry growth rates of 35–50 per cent per annum, and worldwide markets of $100–150 billion by the year 2000.

Whilst these industry prognostications are impressive, they are often difficult to interpret because of the diversity of both definitions and potential applications for biotechnology. Japanese studies, for example, usually include traditional applications of biotechnology such as food fermentations and antibiotics production, both existing multi-billion dollar markets in Japan, as well as novel applications of bioprocesses. The US Department of Commerce – and consequently most US market analysts – uses a narrower definition: products and processes deriving from genetic or cellular manipulation, microbial technology (but excluding traditional fermentations for foods and beverages), enzyme technology and bioprocess engineering (see Chapter 4). The situation in Europe is even more confusing: each national government produces its own definition of its existing and future bioindustry, usually reflecting the industrial strengths of the country, and the EEC's definitions and guidelines tend to favour political necessity over commercial reality.

The difficulties in defining biotechnology markets stem in part from the enormous range of potential applications for the technology and the relatively recent scientific discoveries that make these applications possible. By almost any definition, biotechnology comprises an array of technologies that can be applied to create or improve products and processes for a spectrum of markets, each driven by its own economic and social forces. Table 10.1 sets out the major markets for biotechnology in this century and attempts to provide a framework that indicates the applications of both 'old' and 'new' biotechnology in these markets. The sections below first analyse some general trends in the commercial application of biotechnology, then assess biotechnology's role in the growth of several key industrial markets.

10.2   Overview of Market Trends

Perhaps the best indicator of the commercial potential of biotechnology is its ability to stimulate investment in research and development. In fact, R & D can be said to be the largest current market for biotechnology. For example, US Department of Commerce estimates put 1986 investment in commercial biotechnology R & D at $1.4 billion, supplemented by over $2 billion in federal funding of basic research in biotechnology. A recent survey of more than 300 Japanese companies put their investment in biotechnology R & D at nearly $1 billion in 1986, with government programmes providing at least an additional $300 million. European statistics are not so readily available, since only a handful of countries publish data on this sector, but indications are that total European investment is probably about half that of the USA.

Thus, the worldwide 'market' for industrial biotechnology R & D exceeded $3 billion in 1986, with much of the spending on projects in the early stages of the commercialisation pipeline. This enormous investment in R & D can pay off only if biotechnology fulfils its promise of producing novel, high-value solutions to industrial problems, and this is the driving force for development of biotechnology markets.

Optimistic estimates indicate the worldwide value of biotechnology products could climb to $150 billion by the year 2000;

196

TABLE 10.1 TRADITIONAL AND NEW BIOTECHNOLOGY APPLICATIONS AND MARKETS

Market sector	Traditional biotechnology applications	New biotechnology applications
Agriculture		
Animal health & improvement	vaccines, animal breeding	vaccines, immunostimulants, embryo transfer
Animal growth promotion	antibiotics, steroid hormones	probiotics, peptide hormones
Plant protection	biological pest control	genetic engineering of pest & disease resistance
Crop improvement	plant breeding	plant cell culture & genetically engineered plants with increased vigour & value
Chemicals		
Bulk chemicals	fermentations of organic acids, organic solvents	improved fermentation economics
Speciality chemicals	fermentations of amino acids, antibiotics	chiral molecules from biotransformations
Enzymes	food processing, beverage clarification	detergents, fatty acid esterification, high fructose corn syrup, chemical degradation
Energy		
Methane production	anaerobic waste digestion	improved yields
Biomass conversion & bioethanol		fermentation from agricultural products
Enhanced oil recovery		microbial products & processes
Environment		
Testing & monitoring	mutagenesis testing	application of diagnostic technology to detection of toxic compounds
Waste treatment	sludge processes, anaerobic digestion	processing & reclamation of industrial wastes
Land detoxification & reclamation		microbial processes that degrade toxic compounds, reintroduction of soil bacteria
Food and beverages		
Testing & monitoring	microbiological testing	application of diagnostic technology
Flavourings & sweeteners	monosodium glutamate	aspartame, improved oils & fats
Preservation	cheese, bean curd, yoghurt	
Nutritional supplements	amino acids, vitamins	single cell protein, mycoprotein
Brewing	beer production	'light' beer
Medicine		
Diagnostics	microbiological testing, clinical chemistry	in vivo immunoassays & DNA probes, in vivo diagnostic imaging
Genetic screening	karyotyping	genetic probes
Therapeutics	antibiotics, other microbial products	biopharmaceuticals, novel microbial products

pessimists predict market values of about one half of this figure. Biotechnology's ability to achieve its maximum or minimum commercial potential will depend not only on its ability to produce innovative products that satisfy market demand, but also on its ability to convince governments and consumers of the value and safety of these products. Factors likely to create opportunities for or threats to industrial exploitation of biotechnology include:

Early success of novel biopharmaceuticals, particularly in treatment of cancer and cardiovascular disease, could reduce regulatory hurdles and speed market growth. For example, the 20 new pharmaceuticals approved by the US Food and Drug Administration (FDA) in 1986 required an average of 34 months of FDA review and pre-approval testing; the four biotechnology-produced products, Orthoclone OKT3, Roferon A, Intron A and Recombivax HB, required only 27, 19.5, 11 and 6 months, respectively. However, some biopharmaceuticals present great difficulties in administration and have side effects that may limit their markets to extreme clinical situations.

Development of biotechnology products is outstripping governments' ability to regulate introduction of these products, most notably in the area of environmental release of genetically modified organisms. Neither the EEC nor the USA has been able to develop a consensus policy on this issue, and legal challenges to field tests have already delayed some agricultural biotechnology projects by years. Much of the potential of biotechnology will remain unrealised unless governments institute decisive, rational policies that both allay public concern and permit controlled testing of genetically modified organisms in the field.

Policy changes that reduce international trade barriers and protectionist legislation are important for growth. The markets for most biotechnology products are international, but national regulatory and economic restrictions often reduce the commercial viability of novel products. 1986 brought several positive developments: new EEC legislation that reduces artifically high prices for fermentation feedstocks, reduction in restrictive regulatory legislation for pharmaceuticals in Japan, and new drug export legislation that will allow US companies to export products that have not yet received FDA approval. This trend must continue and expand for successful exploitation of biotech products.

Political and social policies both limit the economic applications of biotechnology and create market opportunities for development of 'uneconomic' processes. The staggering proportions of agricultural surpluses in the EEC and USA have played havoc with agricultural economics, and made the consumers who ultimately pay for agricultural subsidies question the wisdom of introducing new technologies that further improve agricultural and dairy yields. At the same time, the EEC is actively seeking biotechnological solutions that reduce its grain mountains and wine lakes by converting them to industrial alcohol and other chemical products, processes that could prove economic in the long term but will create yet another layer of subsidised agroindustry in the near term. Irrational as these policies may seem, they are unlikely to change significantly until governments accept the need radically to reform current agriculture policies and practices.

Thus, biotechnology markets will operate in a social and economic context clouded with uncertainty. Public enthusiasm about the potential of biotech is tempered with Faustian fears of genetic manipulation, and the political clout of 'technophobic' interest groups could delay some applications of biotechnology for years, or even decades. If, however, the dozen or so products slated to enter the market in the next few years prove successful in both economic and safety terms, the growth of biotechnology into a multi-billion dollar industry is assured.

10.3 Trends in Pharmaceutical Markets

Biotechnology's impact on health care markets is significant and growing: more than twenty therapeutic products had been launched worldwide by the end of 1986, and medical diagnostics accounted for more than half of the US industry's $500 million product sales last year. But industry analysts predict this is just the tip of the iceberg: biotechnology products could account for nearly 5 per cent of international therapeutics markets worth $100 billion, and 15–20 per cent of in vitro diagnostics markets worth $8–9 billion in the early 1990s.

DIAGNOSTICS

Table 10.2 indicates the growth of worldwide markets for in vitro diagnostics: biotechnologies such as monoclonal antibodies and nucleic acid probes are fuelling the growth of key segments such as infectious and immune disease testing, drug monitoring, endocrine testing, and cancer testing. The biotechnological tests are usually simpler and more rapid than conventional assays, and have therefore complemented and enhanced the two prevailing trends in the in vitro diagnostics market: automation and a greater proportion of tests being done on site. Whilst each immunoassay or gene probe test is very specific, all tests of either type work on the same principle and can use similar detection systems; thus, a single automated system can run a battery of tests. Moreover, these tests are exquisitely sensitive, requiring smaller samples and fewer manipulations than chemical and histological methods, and many can be performed in doctors' surgeries or even at home.

TABLE 10.2   WORLDWIDE IN VITRO DIAGNOSTICS MARKETS

Type of test	1985 ($ millions)	1990 ($ millions)	Compound annual growth rate
Clinical chemistry	2700	3700	7%
Infectious disease testing	850	1300	9%
Haematology	650	950	8%
Drug monitoring	340	550	10%
Endocrine testing	350	750	16%
Blood banks/processing	260	450	12%
Immune disease testing	250	450	12%
Cancer testing	100	250	20%
TOTAL WORLDWIDE	5500	8400	9%
Breakdown of infectious disease testing			
Immunoassay/DNA probe			
Gastointestinal	160	300	13%
Sexually transmitted	60	250	33%
Respiratory	40	200	38%
Subtotal	300	800	22%
Conventional/automated microbiology	550	500	−2%
TOTAL MARKET	850	1300	9%

Source: Eberstadt Fleming

The relatively low entry barriers of in vitro diagnostics markets made them an early target for applications of biotechnology, but markets for these products are likely to level off by the mid 1990s as demand for new tests falls and competition drives down prices.

Fortunately, two major markets for the technologies will be emerging by then: genetic screening and in vivo diagnostic imaging.

Genetic defects are linked to many common diseases, e.g. diabetes, cardiovascular disease, cancer and Alzheimer's disease, as well as purely genetic diseases such as cystic fibrosis, Duchenne's muscular dystrophy, Down's syndrome and a variety of blood disorders. Current genetic screening tests involve laborious chromosome analysis or complicated analysis of DNA polymorphisms; they are expensive and time-consuming and therefore limited to patients at high risk. Tests based on specific genetic probes will be quicker and cheaper, permitting widespread screening of neonates and potential parents in high risk categories, a market worth $400 million per annum if tests cost about $50 each.

Genetic screening for defects linked to more common diseases offers a much larger market: assuming just one test per person lifetime, the US market could be over 3 million tests per annum. Successful development of this market, however, will require careful consideration of the ethical and social issues that attend genetic screening programmes; debate over the potential value and abuses of genetic screening has already begun, and the progress of this debate will determine the real market potential for this technology. The application of monoclonal antibodies to in vivo diagnostic imaging offers a more secure market opportunity, and this sector is likely to represent 20–25 per cent of the $4.5 billion monoclonal antibody market by 1995.

THERAPEUTICS

Whilst the market success of biotechnology-based diagnostics is assured, it is the high risk/high reward therapeutics market that will produce most of biotechnology's blockbuster products over the next 5–10 years. Novel blood proteins top the list: tissue plasminogen activator (tPA) is likely to be marketed in Europe and the USA in the next year, with worldwide sales forecast to climb to at least $900 million by 1990. The US company Genentech was stated to be the first to launch tPA, but its failure to win FDA approval in 1987 has been a major setback. Competition for market share will heat up as Wellcome, Biogen, Damon and Chiron (the latter three in partnership with large pharmaceutical firms) introduce tPA products and 'second generation' thrombolytics are launched.

Erythropoietin (EPO), a hormone that stimulates production of red blood cells, could also receive regulatory approval in 1987. Described by one industry analyst as 'the safest and most efficacious drug in the biotechnology industry', EPO's primary market is for treatment of anaemia associated with kidney dialysis. Amgen, a US biotechnology company, has received orphan drug status for this application, which could command a worldwide market of $225 million by 1990. This estimate could double if EPO proves effective in treating other anaemias and stimulating red cell enrichment of blood for autologous donations prior to elective surgery, an increasingly common practice due to fear of AIDS.

Superoxide dismutase (SOD) is a scavenger enzyme that neutralises superoxides, oxygen free radicals that can cause significant tissue damage when blood flow is restored to previously blocked arteries or organs removed for transplantation; it also has anti-inflammatory activity. Human trials with SOD began last year, and the product is unlikely to reach the market until 1989. But lack of alternative therapeutic options indicates that SOD could quickly capture a market of $300 million for cardiovascular and organ transplant applications.

Anticancer agents and immune modifiers could form the biggest potential market for biotechnology in the long term: 1990 sales could top $2 billion worldwide if several of the products prove themselves. Three alpha-interferons have received regulatory approval, primarily for treatment of hairy cell leukemia, a rare but lethal cancer, and Kaposi's sarcoma, a cancer that afflicts many AIDS patients. Extension of approval to therapeutic applications for more common cancers and treatment of infectious diseases is likely in 1987–1988. Clinical trials with beta- and gamma-interferons for cancer therapy are also producing positive results, particularly in combination therapies that use a mixture of immunomodulators and/or traditional chemotherapeutics to kill tumour cells whilst stimulating the body's natural defense mechanisms. Colony stimulating factors (CSF) activate or promote the growth of several important classes of immune system cells: they can, for example, stimulate the growth of bone marrow cells after transplantation or destructive chemotherapy.

Interleukins and tumour necrosis factor have proved more problematic in early clinical trials, but several recent reports indicate these molecules can be used effectively, although mainly in very radical therapeutic regimens that are unlikely to see wide use. In fact, it will take some years before the clinical potential of most lymphokines and other immune regulators is realised: they are powerful, multifunctional mediators, and the success of clinical trials thus far has been marred by side effects, sometimes severe ones, and limited regression of the targeted diseases. Nevertheless, industry analysts remain sanguine about the prospects for these products because there are few alternatives for treating immunocompromised patients, a population that is growing with the spread of AIDS and increased use of chemotherapy, and predict that US sales could reach $650 million in 1990. Table 3.1 of Chapter 3 lists one top US analyst's predictions of the markets for these drugs, along with his estimate of the probability that each drug will achieve these targets.

Approval in 1986 of the first therapeutic monoclonal antibody, Ortho Pharmaceuticals' Orthoclone OKT3 for prevention of kidney transplant rejection, renewed optimism about the therapeutic potential of this technology. The big markets for monoclonals are antibacterial therapy, and perhaps prophylaxis, particularly for hospital-acquired infections. Monoclonal antibodies are also being developed as 'magic bullets' that can target delivery of drugs to specific cells; e.g. an antibody directed against a tumour-specific antigen can be linked to a cytotoxic drug, vastly increasing its efficacy. Markets for therapeutic monoclonals are estimated to be worth $300 million in 1990, soaring to $1 billion by 1995.

VACCINES

Hormones and vaccines are both markets that have seen the successful introduction of recombinant DNA products that replace traditional ones. Both Merck and SmithKline have launched recombinant hepatitis B vaccines that are cheaper and safer than those derived from blood, and at least four other hepatitis vaccines are well along in the development pipeline. In the West, hepatitis B vaccination is limited to a few high risk groups (certain health care personnel, dialysis and multiply transfused patients, drug abusers and homosexual men), although the new vaccines are likely to increase acceptance of immunisation amongst this population and double the current $60–70 million market by 1990.

The big markets for hepatitis B vaccines are in the Far East, where the disease is endemic and chronic infection is the major cause of liver diseases such as cirrhosis and cancer. But the relatively high cost of the vaccines will limit their markets to a few of the wealthier countries in this region until improved technology brings the inexpensive products necessary for mass immunisation programmes. In fact, virtually all large vaccine markets are tied to national or international public health programmes, few of which have the resources to pay for novel products. Moreover, the economics of vaccine development are clouded by product liability issues: because vaccines are generally given to healthy individuals, the 'acceptable' levels of side effects or adverse reactions are much lower.

Until there is broader medical and public awareness and acceptance of risks associated with preventive medicine, the growth of these markets is likely to be slow. However, the emergence of fatal infectious diseases such as AIDS, for which vaccination offers the best medical strategy for controlling the spread of the disease, has created some high-value market opportunities that are stimulating vaccine R & D, although products are unlikely to emerge for several years.

HORMONES

Eli Lilly's human insulin, the first recombinant DNA product to receive FDA approval, is finally beginning to claim a significant share of the insulin market, and human insulins are likely to account for over 50 per cent of worldwide insulin sales by 1990. Genentech's human growth hormone (hGH) was granted orphan drug status in response to fears arising from reports that linked several cases of a rare viral disease to hGH preparations extracted from human cadavers, and two other recombinant hGH products were recently granted similar status. The market for hGH as an orphan drug is small, but the hormone may also be effective in stimulating growth of children of short stature, an application that could easily stimulate worldwide market growth to $500 million in 1990.

Factor VIII is the blood-clotting factor that Type A haemophiliacs lack. It is currently produced from blood, and Factor VIII preparations contaminated with HIV-1 transmitted AIDS to haemophiliacs. Although blood used in Factor VIII preparations is now screened for AIDS antibodies, fear of infection with this and other viruses has created a market for recombinant Factor VIII. Earliest approval of new Factor VIII products is forecast in

1989–1990, with a potential world market of perhaps $200 million.

Biotechnology is also being applied to produce hormones not currently available. Epidermal growth factor is being evaluated for wound and burn healing and for certain ophthalmic applications. It could reach the market in early 1988, with successful launch bringing annual sales in the USA to $100 million in 1990. With hormones proving to be a market with somewhat lower entry barriers than many other biotechnology therapeutics, numerous companies are targeting these products for development. Competition could become fierce because markets for products such as insulin, hGH and Factor VIII are limited to individuals that cannot produce either active versions or adequate quantities of the products.

CONCLUSION: PHARMACEUTICAL PROSPECTS

Whilst biotechnology will account for just a few per cent of pharmaceutical industry revenues by 1990, its underlying impact on the pharmaceutical industry will be far more profound. Most biotechnology pharmaceuticals fall into the category of truly new products, which while representing just 10 per cent of the new chemical entities introduced each year, produce 50 per cent of the incremental new product revenue of pharmaceutical companies. Moreover, the Pharmaprojects 1986 survey of the leading R & D therapeutic categories put biotechnology at the top of the list, with 557 potential new drugs, more than the next two categories, anti-inflammatory and immunological cancer drugs, combined.

Pharmaprojects also rated seven biotechnology companies among the top 100 pharmaceutical companies in terms of R & D products: Genentech ranks number 37 worldwide, ahead of industry giants such as Glaxo, Du Pont and Green Cross, with 31 R & D drugs; Chiron, a company not much over five years old, ranks 51, with 24 new products; California Biotechnology and Biogen, each with 18 R & D drugs, ranked in the mid 70s; Nova Pharmaceuticals and The Liposome Company have 16 and 15 new products, respectively, and rank in the 80s; Cetus, with 12 R & D drugs, ranks 99.

The US government recently passed fiscal legislation that will boost companies' potential revenues from these R & D products, permitting them to charge for drugs undergoing clinical evaluation. The pharmaceutical industry is, of course, rife with examples of promising R & D drugs that never make it to the market, and it remains to be seen whether biotechnology products will perform better than traditional chemicals in this respect. The success of a few early products, however, has generated real optimism that biotechnology will lead the way to improving the pharmaceutical industry's occasional ability to convert research into profits.

10.4   Trends in Agricultural Markets

A recent survey ventured to guess that biotechnology already has the potential to add $5 billion per year to the value of the world's major crops, a figure that could climb to $20 billion per year by 1995. But in an economic sector dominated by grain mountains, subsidies equating to 5–10 times the market value of crops, and dumping of surpluses disguised as food aid, this potential remains largely unrealised – and in fact seems almost ludicrous. It is agricultural policies, not technologies, however, that have created the current lunacy of agricultural markets, and if new technological developments mean farmers can raise their crops or animals more profitably, then they will continue to demand and use them.

Because politics is likely to continue to be at least as important as supply and demand in the evolution of agricultural markets, most analysts tend to hedge their guesses of the market potential of agricultural biotechnology. Aside from the technical hurdles in this field, which are formidable, two key factors make it unlikely that biotechnology will have a major impact on agricultural markets until well into the 1990s.

First, environmental lobbies have raised public fears over the impact of release of genetically engineered organisms into the environment, fears that are even blocking attempts to assess rationally the potential risks associated with environmental release. Although field tests of several genetically engineered organisms have finally been approved this year, the legal battles to gain these approvals have been costly in both time and money. Without a clear framework for demonstrating the environmental impact of new technologies, industry sources worry it could take a decade to obtain full-scale product approvals. Second, unlike novel pharmaceuticals, which often have an 'instant' market amongst a well defined patient population, innovative agricultural products can take years to achieve market penetration.

LIVESTOCK

Biotechnology's most immediate impact has been on livestock animal health: markets for animal health products are determined largely by the economics of acceptable losses in livestock herds rather than government price supports, and have therefore been more rational targets for biotechnology products. Recombinant DNA vaccines against livestock diseases have been on the market for several years and others are being developed against economically important diseases such as foot and mouth disease, bovine rabies and several insect-borne viruses. Livestock vaccine markets are tough, however, because the economics of livestock preventive health care are marginal, although vaccine demand surges in the face of a potential epidemic. The advent of inexpensive diagnostics for major livestock diseases, which permit early detection of potential epidemics, is fuelling demand for improved vaccines.

Because much of a farmer's investment in livestock comes near the end of its life, i.e. in fattening and shipping, treatment of diseases associated with factory farming methods offer more lucrative potential markets. Several companies have developed bovine lymphokine cocktails (BLC) for treatment of shipping fever, which currently is responsible for annual losses of $3 billion and could command a $50 million market in the USA by 1990. Because European livestock farming and processing are more dispersed, shipping fever is less of a problem, but the general immunostimulatory effects of BLC probably help protect the animals from a variety of infections, and as improved technology reduces the price, BLC markets will expand.

The really large opportunities in animal agriculture lie in growth promotants and feed additives, markets where traditional biotechnology has already had considerable success. In 1985 European farmers spent about $364 million on feed additives such as amino acids ($95 million), growth promotants (hormones and antibiotics, $85 million) and anti-disease agents (antibiotics and vaccines, $62 million), and markets for these products will exceed $400 million by 1990 as biotechnology increases both the production and variety of these products. But increased consumer concern about the long-term effect of feed additives is already having radical effects on these markets.

The 1986 decision by the EC Council of Ministers to ban five hormones used to promote growth in beef cattle is a case in point.

After several years of study, a scientific committee appointed by the Commission concluded use of the hormones presented no danger to consumers, but emotional arguments, and perhaps covert protectionism (since the ban could effectively block US exports to the EEC worth $100 million a year) won the day, although banning the hormones will do little to reduce surplus beef mountains.

Several new genetically engineered peptide growth hormones are slated for launch next year. These peptide hormones are quite different from the steroid hormones currently available; their activity is much more specific and results in significantly improved feed conversion rates, increasing milk production by as much as 20 per cent in dairy cows and promoting growth of lean tissue in beef cattle and pigs. Moreover, the growth hormones are broken down in the stomach, so even if they do end up in milk or meat the chances of them affecting consumers are very slim. The prospects for these hormones are good in the meat-loving US market, where consumer lobbyist complaints about factory farming techniques largely fall on deaf ears. In Europe, controversy has already erupted over trials of bovine somatotrophin (BST) on dairy cows, and opposition to registration of growth hormones for livestock is likely to be strong. Thus, while analysts predict that BST's ability to improve dairy yields and produce lean beef make it a potential billion-dollar-per-year product, few of them will bet on when, or even if, it will achieve that potential.

Concern about the use of antibiotics in animal feed has a more rational basis: development of antibiotic-resistant organisms has been fairly convincingly traced to feedstuffs mistakenly overdosed with antibiotics. Whilst there is not yet strong evidence for normal levels of antibiotic feed additives inducing resistant strains, antibiotic resistance is a significant problem when it occurs because it is carried by small genetic elements called plasmids that can spread quickly amongst different bacterial species. This has created demand for alternative strategies to deal with the stress-related problems that antibiotics now solve. Microbes may again provide the solution, this time in the form of probiotics, mixtures of bacteria that can be fed to animals to help restore the chemical imbalances caused by stress. The UK Ministry of Agriculture approved the use of probiotics in animal feed last year, but their use is not yet widespread enough to prove their long-term value – or to generate consumer protests. Probiotics are used successfully in Japan, particularly in pigs, and could represent a significant new animal health market for biotechnology.

The scope for genetic improvement of most Western livestock is limited: conventional breeding has already produced most of the achievable gains in productivity or adaptation to local conditions. But this is not true in the Third World, where local breeding stock is often of poor quality and champion stock are ill-equipped to deal with local conditions. Embryo transfer can speed the introduction of new breeding stock enormously. Embryos from genetically favoured parents can be artificially implanted in surrogate mothers from local stock, where they acquire some of the surrogate's immunity to disease whilst developing into genetically superior calves. Although this technology remains relatively expensive, $75–100 per transplant, it is already being used to improve breeding stock in Latin America and some Asian countries, and continued technical improvements will drive down the price.

The long-term economic effects of this technology could be profound, producing both improved nutrition for the Third World and international competition in agricultural markets. Moreover, development of animal product self-sufficiency in the Third World could be the final blow to the subsidised agricultural systems of the West, although the outcome of such a restructuring is impossible to predict.

CROPS

New developments in biotechnology for crop agriculture may have even more disruptive social and economic implications for international agricultural markets, although the impact of these developments will have to wait until technical and regulatory hurdles have been overcome. The current range of biotechnology products has been produced by non-genetic techniques that speed up conventional crop improvement programmes but are limited to exploiting the existing genetic capacity of the organisms; thus they do not differ significantly from products produced by traditional technologies, and their market impact has so far been negligible.

The big gains in this field will come from selectively introducing valuable new traits into plants or soil organisms: increased vigour and yield, resistance to diseases or pests, value-added traits such as improved nutritional value or flavour characteristics, and resistance to agrochemicals such as herbicides and pesticides. For example, fungal resistance could add as much as 12 per cent to the value of certain seeds (Table 10.3) and a package of herbicide and herbicide-resistant seed can command a 10 per cent price premium.

TABLE 10.3   MARKET POTENTIAL OF FUNGAL-RESISTANT SEED

Crop	Current value ($ million)	Added value of fungal resistance ($ thousand)
Corn	500	200
Cotton	50	300
Peanuts	50	6000
Soya beans	300	200
Tobacco	75	100
Wheat	300	200

Source: *Technical Insights*

CONCLUSION: AGRICULTURE PROSPECTS

In the longer term, agricultural biotechnology must begin to address the fundamental problems of world agricultural markets. Developed countries in Europe and North America want to maintain their agricultural heritage, but can do so in an economically rational way only by replacing surplus crops such as wheat with, for example, new varieties of edible oil plants with improved flavour characteristics. The developing world needs crops that alleviate rather than exacerbate the problems of poor farmers who neither need nor can afford high input farming practices. But the markets that will drive these developments cannot assert their pull until the leviathan of agricultural subsidies and protectionist policies is slain.

10.5   Trends in Industrial Markets

ENZYMES

Industrial enzymes are one of the best established 'traditional' biotechnology markets, currently worth more than $400 million worldwide. Food processing enzymes currently account for about two-thirds of this market: amylases that convert starch to glucose, proteases used in cheese making and to tenderise meat, glucose isomerase for high fructose corn syrup production. The 16 bulk enzymes used in industrial processes accounted for about 90 per cent of the 1985 enzyme market. Whilst bulk enzyme markets will continue to grow at a respectable 7 per cent per annum through to 1995, rapid expansion of applications for high value/low volume speciality and biomedical enzymes will drive growth, and this sector will represent about 45 per cent of the $800–$900 million enzymes market in 1995.

Three key developments will fuel the growth of enzyme markets:
- many new enzymes will be introduced to the market, particularly as diagnostics and treatments for human and animal diseases;
- genetic engineering and improved fermentation technology will both increase the range of enzymes available for industrial use and reduce the costs of producing them;
- new and improved technologies for immobilising enzymes, which increase reaction rates and permit continuous processes, will extend the industrial applicability of enzymatic catalysis.

Thus, biomedical enzyme markets will grow by an average of more than 25 per cent per annum for the next 8–10 years and will represent the largest sector of the enzyme market by 1995, and much of the growth in non-biomedical sectors will come from immobilised enzymes that can be used in new industrial processes. Some processes also require several enzymes, or enzymes that cannot be isolated readily, and systems that use immobilised whole cells or cell components will extend the range of biocatalysis.

FERMENTATION

Most markets for chemicals produced by fermentation are fairly mature, and new technologies have yet to provide the breakthroughs that will make fermentation an economic process for production of many bulk chemicals. Markets for steroids and antibiotics are predicted to double over the next twenty years, with organic acids growing more slowly. Two areas where improved fermentation economics could create significant market opportunities are biopesticides and biopolymers.

Biopesticides currently represent less than 1 per cent of the pesticide market, but environmental concerns are fuelling demand, and development of cheap bulk production processes could drive this up to 20 per cent by 2000. Biopolymers already command significant sales for high value industrial and food applications, but market growth has been limited by fermentation economics. With improved process control and yields, markets for speciality biopolymers such as surfactants, adhesives, fibres and gums could grow to $300 million over the next few years.

Whilst the EEC has not yet abandoned all its aspirations to produce fuel alcohol from its grain mountains, the consistent failure of the programme's supporters to develop it under any rational economic framework has made its future doubtful. EEC efforts to creat new biotechnology-based agroindustries are now concentrating on more realistic goals of developing new crops that could be the source of high-value chemical intermediates such as fatty acids. Several crops high in oleic or erucic acid have been targeted, and whilst the economic wisdom of such a strategy has yet to be proven, it represents a more rational alternative than most of those pursued previously by the EEC.

10.6 Long-term Prospects

As biotechnology moves from research to manufacturing, it is creating new markets for the processes and products that produce and purify new bioproducts. US companies have already invested more than $5 billion to scale up biotechnology processes, and estimates put worldwide investment at nearly $9 billion in the 1986–1990 period. European companies have lagged behind the USA and Japan in bioproduct development, thus growth of European markets for process plant equipment such as fermentors, bioreactors and downstream processing equipment will not surge until the 1990s. Moreover, European users will continue to be concentrated in traditional industrial sectors such as chemical, food and beverage producers rather than biopharmaceuticals. Table 10.4 outlines growth predictions for biotechnology equipment and scale-up markets. Separation and purification can account for more than 90 per cent of the production costs of bioproducts, and sales of equipment and supplies in this field are growing by about 20 per cent per annum. Filtration markets, which already account for about one-third of sales, will triple by

1990; other key growth areas are chromatography and cell culture products.

Whilst several European companies are international leaders in producing equipment and products for biotechnology, European markets for biotechnology production continue to lag well behind those of the USA and Japan. Europe's strengths in traditional biotechnology areas such as enzymes and some fermentation chemicals are rapidly being eroded as new technologies developed elsewhere make traditional processes obsolete. These trends do not bode well for the future of commercial biotechnology in Europe; European public and private sector investment in biotechnology is well under half that of the USA or Japan, and European industry has been particularly sluggish in responding to the challenges new biotechnologies offer to its markets.

Biotechnology is emerging as one of the key industrial growth sectors of the 1990s, but unless European industries and governments make a concerted effort to exploit biotechnology in a commercially sensible way, European economies will enjoy the benefits of biotechnology only second or third hand, as Japan and the USA forge ahead. In the longer term, commercialisation of bioproducts promises (or threatens) to alter the balance of world power, as the Third World begins to benefit from the improved health care, nutrition, industrial advances and plentiful energy sources which biotechnology has the potential to provide.

TABLE 10.4 BIOTECHNOLOGY SCALE-UP MARKETS TO 2000
(figures in $ millions)

Sector	1986–1990 US	1986–1990 World total	1991–1995 US	1991–1995 World total	1996–2000 US	1996–2000 World total	Average annual growth rate
Biopharmaceuticals	1410	5950	2169	9155	3338	14086	9%
Speciality chemicals	410	1750	631	2693	971	4143	9%
Agricultural/other	180	800	275	1231	426	1894	9%
Totals	2000	8500	3075	13079	4735	20123	

Source: Business Communications Corp.

Part Four
Information Resources for Biotechnology

Information Resources for
Biotechnology

GUIDE TO SOURCES

The purpose of this section is to point the reader to further sources of information concerning biotechnology, and to discuss ways of managing the information explosion which has hit this and so many other high technology areas. Biological processes to degrade the lignocelluloses which give wood its structural rigidity are a serious research topic, but a cynic might well argue that the consumption of wood for paper is already well controlled by biotechnology publishing! The last ten years have seen a considerable but largely necessary increase in the amount of information available on biotechnology, particularly concerning the business aspects.

The 'Source Lists' below list numerous sources for further reading and research. A more detailed analysis of what is available is given in 'Information Sources in Biotechnology', A. Crafts-Lighty, Macmillan/Stockton Press, 1986 (second edition).

THE PRESS

Everyone gets both too much and not enough information. It is in keeping up to date that the overload problem is most quickly felt. There are over 40 newsletters specialising in the biotechnology industry. There are maybe a dozen trade magazines for any application sector (pharmaceuticals, chemicals, etc.). Many scientists will need to read at least five 'core journals', and relevant articles could appear in hundreds of other technical publications.

Source List 1 lists some current awareness magazines which are particularly popular in the biotechnology industry. This is by no

217

means a comprehensive list of good publications, but merely a selection covering a broad range of topics and illustrating a wide range of journalistic styles, publication frequencies and subscription prices. Many more general business publications also carry some news on this industry (*Business Week, The Economist, The Financial Times, Fortune, The Wall Street Journal,* etc.).

Fortunately, for biotechnology business information there is a cost-effective monitoring service to supplement and index the primary news sources: *Abstracts in BioCommerce (ABC)*. *ABC* is a twice monthly publication which provides brief summaries of relevant articles plus references to enable you to obtain the original item. *ABC* abstracts appear two to three weeks after the initial publication of the articles, so it is not quite as up-to-date as reading everything yourself but that is an almost impossible prospect. *ABC* is much faster than traditional scientific abstracting operations which can take 3–12 months to index an article. *ABC* covers a wide range of English language sources from all over the world, including some newspapers, and is automatically despatched by airmail. A subscription includes four quarterly cumulative reference editions indexed by organisation. For subject searches or retrospective checks through the 90,000 articles *ABC* has indexed since 1981, telephone access to a computer database version is available through the 'host services' Data-Star and Dialog (see below).

Keeping up to date with scientific information is both harder and easier than business information. It is harder because there is more of it, but it is easier to handle because 'secondary sources' (indexes in the form of abstracting journals and databases) are well developed. Key titles for genetic engineering include *The EMBO Journal, Gene, Nature, Nucleic Acids Research, Plasmid,* Proceedings of the National Academy of Sciences (USA) and *Science.* Immunologists will want some of those plus the *Journal of Immunology* and other publications. Clinicians generally read *The Lancet* and the *New England Journal of Medicine.*

The problem with scientific abstracts is that most of them are not very current, concentrating more on comprehensive coverage. A notable exception is Derwent Publications' *Biotechnology Abstracts*, a twice monthly bulletin covering scientific journals and patents. It is an ideal complementary source to *ABC*. Patent coverage is restricted to the first publication in major countries and is somewhat slower (12+ weeks) than Derwent's main patent indexing service, the World Patents Index.

ELECTRONIC DATABASES

Publicly available databases are produced by a variety of publishers, most of whom are companies already well established in traditional 'print' publishing. A few are government funded or non-profit organisations (such as the US National Library of Medicine and CAB International). Most of the databases are based on printed scientific abstracting services covering scientific and technical literature, but business information is a large growth area. There are several databases providing company profiles including financial data on public companies (e.g. *Disclosure*). Some market information is also available (in abstracts of trade oriented publications such as those covered by *Abstracts in BioCommerce* for example), and some databases listing published market surveys exist.

A current awareness function can be provided by regular searching of the data in the latest update to the database. Most are added to monthly. A search strategy 'profile' may be stored and rerun at intervals, automatically providing a selective dissemination of information (SDI) service. Some of the best databases for biotechnology are listed in Source List 2. Again, this is not a comprehensive list but rather a good starting point. Most databases are sold through agents, 'hosts', also known as 'spinners', who have large computers on which up to 350 different databases are stored. Some of the most important hosts include BRS, Data-Star, Dialog, ESA-IRS, Pergamon-Infoline and Telesystems. At present, Dialog, a US vendor based in California, dominates the market, but some of the other services offer exclusive files or lower rates.

To use a database host, 'network user identification' code (NUI) for your national 'packet switching service (PSS) network, is necessary allowing connection to the international network if necessary.) To get started, contact one or more hosts (see Source List 2), selecting these on the basis of which databases they offer. They will be able to give you more details on how to obtain a PSS NUI and can also advise on modem requirements. Passwords are issued to control access and billing and usually require the completion of a formality contract accepting responsibility for bills but not necessarily committing you to any guaranteed level of usage (discounts may be available if you do).

Searching involves using the host's retrieval software, which will have a specific syntax. For example, on Dialog, in the BioCommerce Abstracts (file r286), 'S *Diagnostic*(W)*test?*' would retrieve all abstracts containing the phrase diagnostic test. 'S *Diagnos?*(w)*test?*' would broaden this to include '*diagnotic tests*', '*diagnostic testing*', '*diagnosis testing etc*'. 'S *Diagnosis?* and *test?*' would get any abstract with both those words in it, say in a sentence like 'The test is used to diagnose . . .'

If you need to search many databases regularly on several hosts, it is best to employ an information scientist trained to do this. For very occasional searching, various commercial 'broker' services are available. In the UK a good low cost service of this type is provided by the British Library's Science Reference Information Service (SRIS) in London, but many freelance consultants will also do this for you.

BOOKS

The European Biotechnology Information Project (EBIP) produces a low cost bulletin listing new biotechnology books and conference proceedings, available at the British Library's Science Reference Information Service (SRIS) and this may shortly be available on a commercial basis. Most of the biotechnology books available, however, are technical monographs or papers published after a scientific meeting. When it comes to buying biotechnology books, generally you will not find many of those in Source List 1 in a bookshop. Even large university bookshops will carry only a few textbooks and some current technical monographs in stock for browsing. Many publishers will accept direct orders by mail, or a specialist purchasing agent may be used. Most will be happy to announce new books.

CONFERENCES AND EXHIBITIONS

The biotechnology field is a good one for conferences. Naturally, there are many scientific conferences which have developed from the long-standing tradition for regular, often international, meetings for biochemists, chemists, biologists and doctors. Many of these are organised by scientific or professional societies on a breakeven basis financially. They are hence quite inexpensive to attend but very technical in contact. Some of the key papers are summarised in the biotechnology newsletters and often the proceedings are published as a book. It is worth noting that such proceedings usually don't come out until a year or so after the conference. In biotechnology, there is much less demand in the mid 1980s for general briefing meetings and big conferences with a very broad concurrent programme. The trend is mainly towards smaller, more specialist events. Many conferences also include an exhibition. Shows are used mainly by companies selling reagents and services, publications and equipment *to* this industry rather than by the specialist biotechnology companies themselves. To find out what conferences are on, the best first step is listings and advertisements in magazines read for current awareness. Many companies find it useful to produce an internal listing for employees' use, compiled after scanning all announcements received and magazine listings, and the British Library's European Biotechnology Information Project (EBIP) produces a monthly conference bulletin available on subscription.

MARKET AND COMPANY INFORMATION

Market information is a very difficult but important area for biotechnology. Over 400 surveys have been published in this field, ranging in price from a few hundred dollars to over $30,000 for a multi-client study over several years. Most cost around $1000–$5000. A listing of those available is given in the UN Department of Trade's 'Marketsearch', and again EBIP produces a bulletin periodically listing recent surveys, although most are too expensive for the British Library to have reference copies. Source List 4 contains a selected list of reports published in 1986, including some financial reports on key companies.

Some specialised requirements will, of course, require privately commissioned research, and very broad interests may be best served by large multi-client studies run over several years. Well-known firms providing that type of project include Arther D Little and SRI International. However, if your resources don't run to the $20,000+ price of such studies or even the more modestly priced reports listed in Source List 4, a great deal of market information appears in the trade literature.

Some market and technical data are fairly static and hence can be found in reference works. Trade statistics are usually too broad in their categories to be useful, but medical and agricultural statistics can be helpful (although most countries' figures are 1–3 years out of date when published). There are also a few handbooks helpful to biochemists and pharmacopoeia, and directories of research institutes will be useful in some cases. It is company directories which have been the most popular new reference work for publishers. The current offerings range from about £50 to over £500 and from a few hundred to about 3000 entries. All contain address details and a rough guide to areas of business. Some contain profiles up to 2 pages long. Source List 5 lists some of the most recently updated directories. Only some are indexed in any way – generally they are best used to get a quick background on a specific firm.

So many biotechnology companies have now gone public that getting financial details has become much easier in the last five years. In the UK, all company results are available from Companies House directly or through agents. Financial results are also released to the press and so will be picked up along with other business information in magazines and databases. Sometimes some details such as turnover will also be included in printed directories.

Source List 1 – Useful Newsletters and Magazines Covering Biotechnology

Title	Publisher	Frequency[1]	Focus
Agrow	George Street Publications	TM	Agriculture
Animal Pharm	V & O Publications	TM	Veterinary
Bio/Technology	Nature Publishing Co.	M	Biotechnology
Biotechnology Newswatch	McGraw-Hill	TM	Biotechnology
Clinica	George Street Publications	W	Diagnostics
European Chemical News	Industrial Press	W	Chemicals
Genetic Engineering News	Mary Ann Liebert	B	Biotechnology
Nature	Macmillan Journals	W	Molecular biology
Science	AAAS	W	Molecular biology
Scrip	PJB Publications	TW	Pharmaceuticals

Note:[1] M = monthly, B = bimonthly, TM = twice monthly, W = weekly, TW = twice weekly

Source List 2 – Major Online Hosts & Databases

HOSTS

BRS (Bibliographic Retrieval Services Inc.) – 1200 Route 7, Latham NY 12100, USA
Data-Star – Plaza Suite, 114 Jermyn Street, London SW1Y 6HJ, UK
Dialog Information Retrieval Service – 3460 Hillview Avenue, Palo Alto CA 94304, USA
ESA-IRS (European Space Agency Information Retrieval Service) – ESRIN, via Galileo Galilei 00044 Frasiati, Italy
Pergamon-Infoline Ltd – 12 Vandy Street, London EC2A 2DE, UK
Telesystems-Questel – 40 rue de Cheche-Midi, F-75006 Paris, France

DATABASES

Name	Hosts
Abstracts in BioCommerce (BioCommerce Abstracts)	Data-Star Dialog
Biotechnology Abstracts	Pergamon-Infoline
Chemical Abstracts	Data-Star Dialog ESA-IRS
Life Sciences Collection	Dialog
Medline	Data-Star Dialog

Source List 3 – General and Introductory Biotechnology Books

Algeny, J Rifkin (Viking Press, 1983). A speculative view of the prospects of genetic engineering in higher organisms, which opposes the prospect of manipulating human genes.

Altered Harvests. Agriculture, Genetics and the Fate of the World's Food Supply, J Doyle (Viking Penguin, 1985). Speculation on the impact of plant biotechnology.

Bio-Japan. The Emerging Japanese Challenge in Biotechnology, J Elkington (Oyez Scientific and Technical Services, 1985).

Biofuture. Confronting the Genetic Era, B K Zimmerman (Plenum, 1984).

The Biotechnological Challenge, S Jacobsson, A Jamison and H Rothman (Cambridge University Press, 1986). Focuses on the potential for developing countries.

Biotechnology. The Biological Principles, M D Trevan, S Boffey and P Stanbury (Open University, 1987).

Biotechnology. An Industry Comes of Age, S Olsen (John Wiley, 1986). Proceedings of a conference on safety issues.

Biotechnology. International Trends and Perspectives, A T Bull, G Holt and M D Lilly (Organisation for Economic Cooperation and Development, 1982). An excellent short overview including a glossary and collection of definitions of biotechnology.

Biotechnology. Laboratory to Marketplace (Open University, 1986). Course text for new biotechnology short course PS625 divided into six blocks in two volumes: Biotechnology made Business, The Pharmaceutical Industry, The Chemical and Food Processing Industries, Agriculture, Energy and Environment, Biotechnology and Society. Other support material including four case studies, two basic science modules, a glossary and bibliography is also available and related videos were produced.

Biotechnology. From Microbe to Market, L S Gerlis and V G Daniels (Cambridge Medical Books, 1986).

Biotechnology. A New Industrial Revolution, S Prentis (Orbit Publishing, 1984). An excellent and enthusiastic introduction.

Biotechnology. The Renewable Frontier, D E Koshland (American Association for the Advancement of Science, 1986).

Biotechnology. Strategies for Life, E Antebi and D Fishlock (MIT Press, 1986). Historical illustrated review.

Biotechnology. The University-Industrial Complex, M Kennedy (Yale University Press, 1986).

Biotechnology (Studies in Biology. Number 136), J E Smith (Edward Arnold, 1985). Good short introductory book now in its second edition.

Biotechnology and British Industry, P Dunhill and M Rudd (Science and Engineering Research Council, 1984).

The Biotechnology Business. A Strategic Analysis, P Daly

(Frances Pinter, 1985). Readable though not always quite accurate review of recent developments in commercial biotechnology.

Biotechnology Made Simple (PJB Publications, 1985). Second edition. An expensive but well written fairly technical glossary of 750 terms used in biotechnology plus a 40-page overview.

Broken Code. The Exploitation of DNA (Seirra Club Books, 1985). Advocates a moral dimension to biotechnology.

A Century of DNA. A History of the Discovery of the Structure and Function of the Genetic Substance, F H Portugal and J S Cohen (MIT Press, 1977).

Cloning and the Constitution. An Enquiry into Governmental Policymaking and Genetic Experimentation, I H Carmen (University of Wisconsin Press, 1986).

Commercial Biotechnology. An International Analysis (Office of Technology Assessment, US Department of Commerce, 1984). A historical analysis concerned that Japanese advances may outpace American success.

Declaration of Heretic, J Rifkin (Routledge and Kegan Paul, 1985). Attack on biotechnology from a well-known activist.

Dictionary of Biotechnology, J Coombs (MacMillan, 1985).

A Dictionary of Genetic Engineering, S Oliver and J Ward (Cambridge University Press, 1985).

Discovering DNA. Meditation on Genetics and a History of the Science, N A Tiley (van Nostrand Reinholt, 1983). A general historical and philosophical treatment. Reproduces some key papers and extracts from private correspondence.

DNA for Beginners, I Rosenfield, E Ziff and B van Loon (W W Norton, 1983).

The DNA Story. A Documentary History of Gene Cloning, J D Watson and J Tooze (W H Freeman, 1981). Large illustrated history including press cuttings.

Future Man, B Stapleford (Granada, 1984). Interesting but extreme illustrated speculations on the potential of human genetic engineering by a well-known science fiction writer.

The Gene Age. Genetic Engineering and the Next Industrial Revolution, E J Sylvester and L C Klotz (Charles Scribner's Sons, 1984).

The Gene Business. Who Should Control Biotechnology? (Crucible Science in Society), E Yoxen (Pan Books Series and Channel Four Television Company, 1983). Based on a television programme. Contends that biotechnology should not be a source of profit.

Gene Cloning. The Mechanics of DNA Replication, D M Glover (Chapman and Hall, 1984). Good technical overview.
The Gene Doctors, Y Baskin (William Morrow, 1984).
The Gene Factory. Inside the Biotechnology Business, J Elkington (Century, 1985).
Genetic Alchemy. The Social History of the Recombinant DNA Controversy, S Krimsky (MIT Press, 1982).
Genetic Engineering in Higher Organisms (Studies in Biology Number 162), J R Warr (Edward Arnold, 1984).
Genetic Prophecy. Beyond the Double Helix, Z Harsanyi and R Hutton (Granada, 1982).
Hormones. The Messengers of Life, L Crapo (W H Freeman, 1985).
The Impacts of Applied Genetics, Microorganisms, Plants and Animals (Office of Technology Assessment, US Department of Commerce, 1981). Slightly dated but excellent overview.
Industrial Biotechnology in Europe, D Davies (ed.) (Frances Pinter, 1986).
Industrial Microbiology and the Advent of Genetic Engineering (W H Freeman, 1982). Reprint of the September 1981 Industrial Microbiology issue of *Scientific American* Volume 245 Number 3. Contains a good collection of fairly technical and slightly dated reviews.
The Interferon Crusade, S Panem (Brookings Institution, 1984). Historical review highlighting policy issues for research funding.
Man Made Life. A Genetic Engineering Primer, J Cherfas (Basil Blackwell, 1982). An exciting, positive and largely popular account.
The New Biotechnology. European Governments in Search of a Strategy, M Sharp (Science Policy Research Unit, University of Sussex, 1985).
Outline of Biotechnology in the UK (Industrial Aids Ltd, 1986). (Fourth edition.) A rather expensive introductory book.
Playing God. Genetic Engineering and the Manipulation of Life, J Goodfield (Harper Colophon Books, 1977).
Principles of Gene Manipulation. An Introduction to Genetic Engineering (Studies in Microbiology. Volume 2), R W Old and S B Primrose (Blackwell Scientific, 1985). (Third edition.) Good technical introduction.
The Politics of Uncertainty, D Bennett, P Glasner and D Travis (Routledge and Kegan Paul, 1986).
A Realistic View of Biotechnology, E H Houwink (ed.)

(DECHEMA, 1984). Published for the European Federation of Biotechnology.

Reshaping Life. Key Issues in Genetic Engineering, G V Nossal (Cambridge University, 1985).

Setting Genes to Work, S Yanchinski (Viking Press, 1985).

The Stellar Thread. A Story of DNA, Evolution and the Immortality of Ideas (American Chemical Society, 1985).

Understanding DNA and Gene Cloning. A Guide for the Curious, K Dilica (John Wiley, 1984).

Unnatural Selection. Coming to Terms with the New Genetics, E Yoxen (Heinemann, 1986).

What is Biotechnology? (Industrial Biotechnology Association, 1984). Excellent 20-page free introductory pamphlet.

Note: Macmillan Press titles are published by the Stockton Press in the USA.

Source List 4 – Market Surveys in Biotechnology, 1986

Abbott Laboratories (Frost and Sullivan, 1986) 16pp, $190. Wall Street Report H-713.

All the Way with TP-A: New Thrombolytic Therapies to Treat Heart Attacks (Kidder Peabody, 1986) $190.

Alternate Site Clinical Testing (Biomedical Business International, 1986) 50pp, $750.

Amgen (Frost and Sullivan, 1986) 22pp, $230. Wall Street Report H-687.

Amino Acids and Small Polypeptides (Business Communications, 1986) $750. Report C-056.

Animal Health – Pharmaceuticals, Biologicals and Nutritionals/ Feed Supplements Markets in the US (Frost and Sullivan, 1986) 300pp, $2750. Report A1588.

Antimicrobials (Frost and Sullivan, 1986) (In preparation). Report E723.

Antiviral Drugs and Vaccines (Frost and Sullivan, 1986) (In preparation). Report E795.

Becton Dickinson (Frost and Sullivan, 1986) 14pp, $185. Wall Street Report H-690.

Biochemical Food Products Market (US) (Frost and Sullivan, 1986) $1700. Report A1506.

Biochemical Process Equipment Market (Frost and Sullivan, 1986) 305pp, $2500. Report E821.

Bioengineered Medical Proteins (Business Communications, 1986) $1750.

Biogen (Frost and Sullivan, 1986) 22pp, $230. Wall Street Report H-686.

Biological Response Modifiers (Theta Corp., 1986) 112pp, $600. Report 625.

Biopesticides (Biotechnology Affiliates, 1986) (In preparation). $1050. Three sections at $450 each.

Biosensors (Theta Corp., 1986) $750. Report 629.

Biosensors and Chemical Sensors (Business Communications, 1986) 212pp, $1950.

Biosensors Market in Western Europe (Frost and Sullivan, 1986) 274pp, $2500. Report E868.

Biosensors: Today's Technology, Tomorrow's Products (Technical Insights, 1986) 160pp, $325.

Bio-Technology General (Frost and Sullivan, 1986) 12pp, $165. Wall Street Report H-689.

Biotech and Speciality Chemicals (Business Communications, 1986) $1950.

Biotechnologies and Bio-Industries in the World (DAFSA, 1986).

Biotechnology and Fatty Acids (Business Communications, 1986) $1950.

Biotechnology and Microbiology in Australia (Science Forum, 1986) $378.

Biotechnology Equipment and Supplies (International Resource Development, 1986) 336pp, $1285.

Biotechnology Impacts on the Food Industry over the Next 15 Years (Bernard Wolnak, 1986) $9150.

Biotechnology in Japan, Israel and Selected European Countries (Theta Corp., 1986) 150pp, $795. Report 630.

Biotechnology Industry (Frost and Sullivan, 1986) 17pp, $1200. Wall Street Report H-646.

Biotechnology Outlook (Frost and Sullivan, 1986) 30pp, $185. Wall Street Report H-715.

Biotechnology Programme (SRI International, 1986/7) $1500. Multi-client programme producing six reports: Biochemical Transformations. Volume 1: Enzymes in Adverse Conditions; Bioreactors and Separation Technology. Volume 5: Mammalian Cell Culture; Diagnostics, An Update Report: Monoclonal Antibodies, DNA Probes and Sensors; Strategy, An Update Report; Regulations, Product Approvals, Biotechnology Update 1986,

Selected Developments and their Commercial Impact, $5000 each.

Biotechnology Produced Therapeutics (Theta Corp., 1986) $795. Report 645.

Biotechnology Proteins (Theta Corp., 1986) $795. Report 636.

Blood Products (Theta Corp., 1986) $750. Report 626.

Business Opportunities in Interleukins, Interferons and Related Products (Technology Management Group, 1986) $995.

Cancer Diagnosis and Monitoring Products (Theta Corp., 1986) $750. Report 608.

Cell Culture Products Market – Europe (Frost and Sullivan, 1986) 300pp, $2250. Report E771.

Centocor (Frost and Sullivan, 1986) 23pp, $275. Wall Street Report H-653.

Clinical Immunoassay Instrumentation (Frost and Sullivan, 1986) (In preparation). Report E862.

Collagen (Frost and Sullivan, 1986) 12pp, $150. Wall Street Report H-666.

Commercial and Industrial Enzymes (Business Communications, 1986) $1950.

Commercial Microbes. A Growing Speciality (Business Communications, 1986) $1950.

The Commercialisation of Biotechnology. From Promise to Profits (Arthur D Little, 1986) $975. In press.

Diagnostic Reagents (Business Communications, 1986) $1950. Report C-060.

Diagnostic Reagents and Instruments: An Overview of the Japanese Market (Robert S First, 1986) $995.

DNA Probe Technology (Frost and Sullivan, 1986) $1900. Report A1479.

DNA Probes in Medicine (Frost and Sullivan, 1986) $1900. Report A1479.

Drug Delivery Systems, Technology, Companies and Market (Biomedical Business International, 1986) 160pp, $1850.

Emerging AIDS Markets: A Worldwide Study of Drugs, Vaccines and Diagnostics (Technology Management Group, 1986) 435pp, $995.

Enzyme Markets (Charles H Kline, 1986).

European Market for Clinical Testing Products (Biomedical Business International, 1986) 80pp, $950.

European Market for Feed Additives: Impact of Biotechnology (Frost and Sullivan, 1986) 255pp, $2350. Report E829.

Genetic Testing USA – 1986 to 1990 (Robert S First, 1986) 100pp, $3000.

Home Diagnostic Testing (Theta Corp., 1986) 122pp, $750. Report 602.

Immobilization Technology (Battelle Columbus, 1986) $32,000. Proposed multi-client survey study C-59.

Immobilized Enzymes, Cells and Bioreactants (Business Communications, 1986) $1950.

Immunodiagnosis Reagents Market in Western Europe (Frost and Sullivan, 1986) 280pp, $2400. Report E861.

Immunodiagnosis Reagents Market – US (Frost and Sullivan, 1986) 200pp, $1975. Report A1626.

Immunodiagnostics USA – 1986 (Robert S First, 1986) $8500.

Immunotherapy Products and Technology Market in the US (Frost and Sullivan, 1986) 208pp, $1850.

The Impact of Biotechnology on Animal Care – A Worldwide Assessment of Opportunities in Diagnostics, Vaccines, Drugs, Growth Enhancement and Other Products (Technology Management Group, 1986) 500pp, $2900.

The Impact of Biotechnology on Animal Feeds and Health Products (Frost and Sullivan, 1986) (In preparation). Report E829.

The Impact of Biotechnology on Pesticides (Frost and Sullivan, 1986) (In preparation). Report 1537.

The In-Vitro Cancer Diagnostics Market in the US (Frost and Sullivan, 1986) 250pp, $1975. Report A1644.

The In-Vitro Diagnostics Market (Frost and Sullivan, 1986) 23pp, $285. Wall Street Report H-716.

Industrial Enzymes: Technology and Markets in Europe (Frost and Sullivan, 1986) 345pp, $2650. Report E814.

Industrial Enzymes Markets (Frost and Sullivan, 1986) 345pp, $1500. Report E814.

Integrated Genetics (Frost and Sullivan, 1986) 23pp, $250. Wall Street Report H-676.

Interferons: Commercial Prospects for the Next Decade (Theta Technology, 1986) 245pp, $795.

Eli Lilly (Frost and Sullivan, 1986) 25pp, $280. Wall Street Report H-659.

Membranes Separations Markets and Technology (Business Communications Co., 1986) $1950. Report P-041U.

Monoclonal Antibodies (Technical Insights, 1986) $645.

New Business Opportunities in Peptides (Strategic Analysis,

1986) $140,000. Three-year programme which will publish some reports in 1986. Multi-client survey. US, Japan or Europe data only available at $1500 each.

New Directions in Diagnostic Reagents (Business Communications, 1986) $1950.

New Horizons in Drugs (Business Communications, 1986) $1950.

New Immobilization Technology. Commercial Opportunities for Reagents, Catalysts and Biocatalysts (Business Communications, 1986) $1950.

Non-Conventional Proteins (Frost and Sullivan, 1986) (In preparation). Report W815.

Opportunities in Viral Testing (Robert S First, 1986) $3500.

Over the Counter Home Healthcare Kits and Devices (Business Communications, 1986) $1750.

Pesticides Market (US) (Frost and Sullivan, 1986) 273pp, $1900. Report A1561.

Product/Asset Valuation: Benchmark for Valuation in Biotechnology (Kidder Peabody, 1986) 30pp, $450.

Protein Engineering: Technical Perspectives and Strategic Issues (SRI International, 1986) 360pp, $5000. (Part of the SRI's 1985/6 Biotechnology Programme now available separately.)

Protein Engineering and Bionetic Engineering. The Next Waves in Biotechnology. A Technical and Commercial Assessment (Strategic Technologies International, 1986) 400pp, $8000.

Protein Ingredients Market (Frost and Sullivan, 1986) $1900. Report A1549.

A Report on the Opportunities for Improving the Potato (John Kitching Associates, 1986) (In preparation). $12,500.

Reproductive Technology Products and Markets (Biomedical Business International, 1986) 120pp, $850.

Research Biochemicals Market in the US (Frost and Sullivan, 1986) 220pp, $1950. Report A1525.

Scale Up in Biotechnology (Business Communications, 1986) $1950. Multi-client survey.

Schering-Plough (Frost and Sullivan, 1986) 14pp, $190. Wall Street Report H-711.

Selling Clinical Diagnostics in Saudi Arabia (Biomedical Business International, 1986) 120pp, $1200. Report C-061.

Separations and Purification Equipment Market in Biotechnology Applications (Frost and Sullivan, 1986) 241pp, $1900. Report A1661.

Separations for Bioprocess (Battelle Columbus, 1986) $32,000. Multi-client survey study C-34.

Thrombolytic Therapies (Frost and Sullivan, 1986) 10pp, $190. Wall Street Report H-706.

Tumours Markers in the Clinical Lab (Theta Technology, 1986) 160pp, $600. Report 540.

The United States Food and Drug Administration Regulatory Process for Medical Devices and In-Vitro Diagnostic Devices (Strategic Technologies International, 1986) 101pp, $400.

The US Clinical Testing Market (Biomedical Business International, 1986) 125pp, $1850.

Veterinary Diagnostic Tests and Instruments (Biomedical Business International, 1986) 80pp, $950.

Veterinary Lab Testing Markets (Theta Technology, 1986) 100+ pp, $795. (In preparation). Report 640.

Veterinary Products Market in Europe (Frost and Sullivan, 1986) 296pp, $2300.

Water Reuse and Recycling Technology (Business Communications, 1986) $1950.

Source List 5 – Recently Published Specialist Biotechnology Directories

Biomass. International Directory of Companies, Products, Processes and Equipment – Macmillan, £45.

BioScan. The Biotechnology Corporate Directory Service – Cetus/Oryx Press, $425.

Directory of British Biotechnology 1987/88 – Longman, £65.

Fifth Annual GEN Directory of Biotechnology Companies – Mary Ann Liebert, $40.

Genetic Engineering and Biotechnology Related Firms Worldwide Directory 1987/88 – Sittig and Noyes, $200.

Genetic Engineering and Biotechnology Yearbook 1986/87 – Elsevier, $800.

Healthcare Biotechnology: Company Profiles – PJB Publications, £75.

International Biotechnology Directory (known as The Biotechnology Directory in North America as published by Stockton Press) Macmillan, £70 ($140).

Source List 6 – Regulations on Biotechnology

Recombinant DNA Safety Considerations. ISBN 92-64-12857-3. OECD, Paris.

Review and Analysis of International Biotechnology Regulations. A D Little. National Technical Information Service, US Department of Commerce.

US Coordinated Framework for the Regulation of Biotechnology. Federal Register 50FR 47174 (14 Nov 1985) and 51FR 23301 (26 June 1986).

Biotechnology: Support and Regulations in the UK. Department of Trade and Industry, London.

Commercial Biotechnology: An International Analysis. Congressional Office of Technology Assessment, Washington. Jan 1984.

The European Community and the Regulation of Biotechnology: An Inventory. Biotechnology Regulation Interservice Committee. Directorate-General for Science Research and Development. 1986.

GMP Guidelines. US Code of Federal Regulations; 21CFR Parts 210, 211 (1985). Also World Health Organisation Reports 'Requirements for Biological Substances: Report of a WHO Expert Group' Tech Rep Ser No 323, 1966; and 'Quality Control of Drugs' WHO, 1977.

Index

forensic testing 36
Foundation for Economic Trends
  153
Fuji 61
Fujisawa 181
fungi 21
Fusarium graminearum 169

gasahol 23
GBF (W. Germany) 120, 121–2
Gen-Probe 191
'gene boutiques' 5
'gene machine' 9
gene-splicing 6, 8
Genetech 31, 53, 62, 64–5, 70, 74,
  115, 126, 130, 131, 179, 181, 182,
  184, 186, 201, 204, 205
  patent on TPA 174
genetic code 8, 10
genetic engineering 6, 8–9, 10, 13, 24
Genetic Manipulation Advisory
  Group (GMAG) (UK) 141, 146
genetic screening 36, 193–4, 201
Genetic Systems 61
Genetica 106, 125, 137
Genetics Institute (Boston) 130, 181,
  183
Genex 62, 68, 70, 115
'germ theory' 4
ginseng 95
Gist-Brocades 106, 132
glandular fever 19
Glaxo 106, 115, 127, 129, 205
gluconic acid 13
glues 24
glycosylation 178
gonorrhoea 190
Good Industrial Large-scale Practice
  (GILSP) 95, 151, 156
Good Manufacturing Practice 142
Grace, W.R. 65
Gram, Hans 16
Green Cross 205
Green Party (W. Germany) 136
growth factors 37
growth hormone 10, 11, 92
  bovine (bGH) 22
  human 58, 84, 94, 131, 174, 179,
    204, 205
Gruenenthal 123
Guinness-Distillers 129

haemophilia 18, 36, 37, 181,
  192, 204

Hanson Laboratories 131
Hayashibara 183
Heineken 132
hepatitis 18, 20, 65, 181, 183
hepatitis B 41, 58, 66, 84, 94, 185,
  203
hepatitis B surface antigen protein
  (HbSAg) 185-6
herpes 65, 186, 191
  swine 58
high fructose corn syrup (HFCS) 16,
  21, 211
Hoeschst AG 106, 109, 111, 117,
  121, 122–3, 126, 133, 137, 181
Hoffman-La Roche 64, 74, 106, 108,
  109, 111, 130, 182, 189
Hormone Research Foundation 174
hormones 13, 37, 178–80
  market for 204–5
  see also growth hormone
'hospital infection' 17
human growth hormone (hGH) 58,
  84, 94, 131, 174, 179, 204, 205
Humulin 179
Huntingdon's chorea 192, 193
hybridomas 19, 21, 33, 187
Hybritech 61, 66, 187
Hybritech Inc. v. *Monoclonal
  Antibodies Inc.* 172–3
ICI 21, 106, 109, 111, 127, 129, 133,
  137
ICN Pharmaceuticals 61
Immunex 62
immunodiagnostics 35–7, 187
Immunologia y Genetica Aplicado
  133
immunomodulators 37
Immunotech SA 126, 127
Imperial 129
Imperial Cancer Research
  Laboratory (UK) 190
Imperial Chemical Industries *see* ICI
in vitro diagnostic imaging 94–5, 200
in vivo diagnostic imaging 36, 190,
  201
INCO 115
Industrial and Technical Research
  Council (Spain) 133
industrial applications 47
  potential, in US 56–7
influenza 183, 185
information resources 217–33
inhibin 186
Institut Merieux 124, 125, 126

Index

Weizmann 5
Wellcome Foundation 106, 111, 115,
    127, 129, 131, 181, 201
white blood cells (B-cells) 19
whooping cough 183
Wistar Institute (Philadelphia) 130
Wistar Institute case (1983) 171–2
World Health Organisation 186
World War I 5

World War II 5, 17
wine 1–5

Xoma 190

yeast 2, 5, 10, 13, 14, 15, 21, 100
    Baker's 5, 9, 34, 178
    Brewer's 2
    'wild' 2